Management for Professionals

For further volumes:
http://www.springer.com/series/10101

Thomas Frisendal

Design Thinking Business Analysis

Business Concept Mapping Applied

Thomas Frisendal
Copenhagen S
Denmark

ISSN 2192-8096 ISSN 2192-810X (electronic)
ISBN 978-3-642-43482-2 ISBN 978-3-642-32844-2 (eBook)
DOI 10.1007/978-3-642-32844-2
Springer Heidelberg New York Dordrecht London

© Springer-Verlag Berlin Heidelberg 2012
Softcover reprint of the hardcover 1st edition 2012
This work is subject to copyright. All rights are reserved by the Publisher, whether the whole or part of the material is concerned, specifically the rights of translation, reprinting, reuse of illustrations, recitation, broadcasting, reproduction on microfilms or in any other physical way, and transmission or information storage and retrieval, electronic adaptation, computer software, or by similar or dissimilar methodology now known or hereafter developed. Exempted from this legal reservation are brief excerpts in connection with reviews or scholarly analysis or material supplied specifically for the purpose of being entered and executed on a computer system, for exclusive use by the purchaser of the work. Duplication of this publication or parts thereof is permitted only under the provisions of the Copyright Law of the Publisher's location, in its current version, and permission for use must always be obtained from Springer. Permissions for use may be obtained through RightsLink at the Copyright Clearance Center. Violations are liable to prosecution under the respective Copyright Law.
The use of general descriptive names, registered names, trademarks, service marks, etc. in this publication does not imply, even in the absence of a specific statement, that such names are exempt from the relevant protective laws and regulations and therefore free for general use.
While the advice and information in this book are believed to be true and accurate at the date of publication, neither the authors nor the editors nor the publisher can accept any legal responsibility for any errors or omissions that may be made. The publisher makes no warranty, express or implied, with respect to the material contained herein.

Springer is part of Springer Science+Business Media (www.springer.com)

Preface

The motivation for this book has been built up over time. Experiences from many different clients in a variety of industries, including the public sector, have made me curious about one simple fact: Business management lost sight of one of its most important assets, the Business Information Asset. This happened during the process of "Management Information Systems (MIS)" changing its name first to "Information Systems (IS)" and then to "Information Technology (IT)." Analysis and design of business information became an engineering style discipline instead of being a business management activity.

Working with business information analysis and modeling (which is what I do) is a very enviable position indeed. When you have the opportunity – as I have – to analyze and design business information in direct dialogue with business people, you get very close to the core of the business; issues which lead directly back to the business model and the business plans.

For several years, the focus for many IT development projects has been business process "renovation." Quite often driven by technology such as object technology leading to service-oriented architectures and the like. However, processes may change quite frequently (and they do), which give them limited long-term business value. Not so for business information, which persist for a very long time and have a high business value.

With this book, I want to help getting focus on business information analysis and design (again). The new approach, I write about, enables business people to work directly and creatively with one of their most important assets – the information about their own business.

What works is a combination of design thinking and concept mapping. Together they enable new, creative conceptual designs and real business innovation. It started around 2004–2005 and has been refined since then. I have had the opportunity to use the concept mapping approach initially on projects together with information management consultants from Devoteam Consulting in Denmark. Since 2007, a Danish business intelligence consultancy with 40+ consultants that I work with on a number of projects, Inspari, has used the approach on assignments for different clients. All together, many business people in both private companies and government have analyzed and designed their business concepts using the method. The results have been excellent, certainly much better than anything else I have seen.

I hope the experiences presented in the book will inspire you and help you to innovate the business of your organization.

There have been a number of people helping me. I am greatly thankful to all the business people, who over the years opened their doors for me and shared their concepts and concerns, and to many consultants at Devoteam and Inspari for excellent teamwork. I am grateful to Prof. Roger Martin of the Rotman School of Management in Toronto for invaluable advice in the early stages of the design of the book. Also, thanks to Mads Carsten Brink Hansen, External Lecturer at The Aarhus School of Business and Social Sciences, Aarhus University, and Business Consultant at Inspari A/S in Denmark for an excellent review. Last, but not least, I am most thankful to my private proofreader, Ellen-Margrethe Soelberg, who also is my patient, loving, and understanding wife.

Copenhagen, 2012   Thomas Frisendal

Acknowledgements

The EU-Rent case study was developed by Model Systems, Ltd., along with several other organizations of the Business Rules Group (www.businessrulesgroup.com), and has been used by many organizations. The body of descriptions and examples may be freely used, providing its source is clearly acknowledged.

CmapTools is a trademark of the Institute for Human and Machine Cognition on behalf of the University of West Florida Board of Pensacola, FL 32501, USA.

Contents

1	**Introduction**	1
Part I	**Design Thinking Business Analysis**	
2	**Understanding the Business**	7
	2.1 Understanding the Business Using Concept Maps	8
	2.2 Information, Data and Business Rules	9
	2.3 Going into Details: Business Rules	10
	2.4 Good Definitions are Important	10
	2.5 What Do Concepts Actually Look Like?	11
3	**Design Thinking for Business Analysis**	15
	3.1 Business Analysis: Understanding Business Information	15
	3.2 Design Thinking: Where Does It Come from?	16
	3.3 Designing Other Things than Products	17
	3.4 Dealing with Wicked Problems	18
	3.5 Design Thinking for Business Development	19
	3.6 Business Synthesis	20
	3.7 Where Does Concept Mapping Come from?	21
	3.8 When Do We Need Tools Like Concept Mapping?	22
4	**Business Analysis Redefined**	25
	4.1 Overview of the Method	25
	4.2 Preparing for the Analysis to Synthesis to Design Flow of Events	26
	4.3 Top Down: The First Workshop	27
	4.4 Explorative Workshops	30
	4.5 Ideation Workshops	31
	4.6 Generalization and Specialization	34
	4.7 Levels of Abstraction	36
	4.8 Implementation Workshops	36
	4.9 Agile Approach	38

Part II Business Concept Mapping

5 Where to Find Meaningful Business Information? 43
    5.1   Start with Your Business Model 43
    5.2   How to Identify Business Concepts 45
    5.3   Concept Mapping Brainstorming Workshops 46
    5.4   Excel: Where Meaning Lives! 48
    5.5   The Chart of Accounts is full of Meaning 53
    5.6   Applications and Databases Might Be Meaningful, Too...... 55
    5.7   Reports 57
    5.8   Documents and the Internet Are Full of Meaning 57
    5.9   Take Control of What Your Business Means! 58
    5.10   Business Dialects 59

6 How to Do Concept Mapping 61
    6.1   Concept Mapping Explained 61
    6.2   What Are Business Objects? 64
    6.3   Properties of Business Objects 64
    6.4   Definitions and Other Specifications 65
    6.5   Structured Concepts 66
    6.6   Layout of Concept Maps 70
    6.7   Dealing with Logic 71
    6.8   When to Stop? 72
    6.9   The Concept Mapping Tool Par Excellence 74
    6.10   Real Life Examples of Concept Maps 74
    6.11   General Company Structure 74
    6.12   Shipping 75
    6.13   Property Management 76
    6.14   Car Dealership 77
    6.15   Public Sector Example 78
    6.16   Concept Harvesting 79
    6.17   Standard Business Concept Definitions 80

Part III Business Innovation Using Mapped Business Concepts

7 Concept Mapping and the Next Generation IT Paradigms 83

8 Opportunity: Reliable Business Information and MDM 87
    8.1   Data Profiling 87
    8.2   Master Data Management (MDM) 88

9 Opportunity: Information Valuation 91

10 Opportunity: Meaningful Business Intelligence 93

11	**Opportunity: Business Rules Automation**	97
	11.1 Concept Maps Versus Business Rules: Revisited	97
	11.2 Business Rules Extend the Concept Maps	98
12	**Opportunity: Reusable Business Information**	101
13	**Opportunity: Open Information Sharing**	107
14	**Opportunity: Pull Instead of Push**	109
15	**Opportunity: NoSQL and Big Data**	111
16	**Think Big, Start Small: Deliver Value to the Business**	113
	16.1 Simple Tools that Work	113
	16.2 Tested Approach	113
	16.3 Benefits of Business Concept Modeling	114
	16.4 Design Thinking	115
	16.5 Summary of Guidelines	117

Appendix 1: Explanation of Terms and Acronyms 119

Appendix 2: User Guide to CmapTools 123

**Appendix 3: Comparison of the Concept Level
and the Entity Level of Modeling** 127

References 133

Introduction

This book is about a new validity-driven approach to business analysis. The approach uses design thinking and concept mapping, which work hand in hand to enable creative, new business concepts and processes. The focus is mostly on the business information side because that is where the long-lasting values of the business model(s) are kept.

The aim of any business development effort is to "tune" the business even better in order to obtain more reliable results. There are two very fundamental requirements to be met by the business analyst/business planner, who wants to be successful:

1. He/she must establish a valid understanding of the business (concepts, and also processes) of the business as it is and as it might be in the future.
2. He/she must be intuitive, insightful and creative during the process as the team learns from the understanding of the business concepts and prototype new ideas in rapid succession.

The valid understanding of the business is delivered by "business concept mapping", a visual, interactive, team-oriented analysis and design method supported by a visual mapping tool. To phrase it in simple terms: This approach is the cheapest way to get real business pains out in the open, and to find ways of doing something about them.

Design thinking for business development learns from the understanding of the business and creates (just like e.g. product designers do) a new or changed set of business concepts, which are valid from the business point of view. Validity and innovation are inseparable.

There is a considerable amount of interest in "design thinking". The context in this book is to facilitate creative thinking within business development. "Design thinking" of business concepts etc. is actually very similar to what designers do. Design is not just for products. *This book adds business analysis to the growing list of design thinking approaches.*

The two approaches (design thinking and concept mapping) work extremely well together. Both of them are based on "learning" and the "psychology of intuition". In combination they have the potential to transform business analysis to become "business synthesis".

Running a business from day to day clearly needs reliability of processes and results. But that does not imply that business analysis should (initially) focus on reliability issues. You cannot deliver reliable results, if your understanding of the business concepts is not valid. Validity is the starting point, and that is what design thinking provides. In order to change something, you must understand it. In most business organizations today there are very optimistic assumptions about the level of understanding of the business concepts among the employees and managers. Incomplete knowledge is a risk to the reliability of the business operations. Concept mapping provides the necessary level of understanding and makes it easy to communicate what "this is all about".

The redefinition of business analysis creates insights within the organization, leading to several potential paradigm shifts to new solution models. Business development is the net result. Some of the major and increasingly popular paradigms in IT are: Information quality and valuation, master data and hierarchy management, business rules automation, business semantics, linked data and lastly the NoSQL and "Big Data" movements. They represent very viable examples of business innovation opportunities using IT technology. The entrance ticket however, is management and design of the core business concepts.

There are three major propositions of this book:
1. When it comes to business concepts and business information, the method of concept mapping effortlessly brings out the best results of design thinking. Concept mapping is an important, new tool for business analysts.
2. The business should – it self – take control of its business concepts and business information. This is a question of business value opportunities and consequently it is a business task, not an information technology task. Business development approaches are increasingly based on design thinking. This agile approach is perfectly supported by concept mapping, a great facilitator for understanding, defining and structuring business concepts and their relationships.
3. There are major opportunities in new business model paradigms and practices, which are only available to those businesses, which manage their business concepts and their relationships.

The key thing to note is that right from the beginning of a business you are basing the business model on a certain set of concepts, definitions and relationships between concepts. Some things may change (e.g. sales channels) but the core of the business model remains conceptually much the same. The result is that businesses, which carefully control, understand and nurture its business concepts, their definitions and structures, and which clearly communicates these matters internally as well as possibly externally, stand to be more successful. From a pragmatic point of view new opportunities present themselves when business concepts are mapped and managed as valuable resources. The business concepts maps become key resources just like the corporate chart of accounts – and they play very similar roles as maps of assets.

1 Introduction

This is a business analysis book. The intended audience includes: business analysts and modelers, business planners, business development teams, enterprise architects, information architects, IT project managers and business managers, of course. The common ground is that the reader should have a business perspective and that he or she wants to help the business develop in creative manners, driven by business information.

Consider this book a guided tour. Our journey will take us through three sections of the new approach to "business synthesis" using design thinking and concept mapping:

1. **Design Thinking Business Analysis** gives you insight into how to deal with what "The Business" is talking about, how to apply design thinking principles for business information analysis and what the redefined flow of the business analysis process looks like.
2. **Business Concept Mapping** takes you into finding meaningful business information concepts, how to do the actual concept mapping and how to "harvest" your business concepts.
3. **Business Innovation Using Mapped Business Concepts** goes – after a short introduction – into different examples of business innovation opportunities. These are only available to business organizations, which manage and design their business concepts. This includes reliability of business information and master data management, valuation of the business information asset, meaningful business intelligence, business rules automation, reusable business information, open information sharing, pull instead of push and lastly the NoSQL/Big Data movement(s).

Business people operate in very dynamic situations and need very flexible approaches. And business people are people, who understand and use different, possibly slightly conflicting contexts, most of the time. We have to *support the reality of business* instead of trying to bend the business towards rigid, engineered and elaborate frameworks and architectures. This is where design thinking comes into the picture. Concept mapping adds its excellent capabilities for understanding, communicating and creating ideas; and for creating those exquisite A-Ha moments!

Remember that business innovation is obviously a business initiative – and the business will support it as long as it adds business value.

Business development is about change and innovation. Before you can change something, you must understand it at a certain level. How, then, do we get to understand the business? This is where we start.

Part I
Design Thinking Business Analysis

"Synchronicity can be explained as an experience, which at times arise in the human brain, as it encounters the disparate events of reality and interprets them as coherent or meaningful. Things and events, which appeared to be random when they occurred, turn up to be part of a greater picture – a structure beneath the skin of the daily life" Martin Bigum, Danish poet/painter, 2012.

Understanding the Business 2

You cannot change something unless you understand it. What can be done to make the reality of the business visible and intuitively easy to understand?

Business starts with ideas. From (some of) the idea(s) a business model is created. The business model is then implemented across and among all the necessary places, things and actors such as people, products, capital, fixed assets, customer segments and so forth. As soon as you are open for business, you start monitoring and you analyze what is going on. Right from the beginning of the business model, you are betting the business on a certain, specified, set of concepts and relationships between them. This is the core of what "the business means". What the business really means is an extremely important issue. Uncertainty – and misunderstandings – on this level can lead to grave problems. The business has a responsibility to be very clear about what it means when it talks about the essential concepts, which are of importance to the business model – and the implementation of it in the daily life of everybody supporting the business.

For some years now many (IT-trained) people have been led to believe that "Use Cases" or similar stories is a good place to start. But how do you write a good case, if you do not know the terminology used by the business? For example: Are we talking about Cars or Vehicles? (And what is the difference between those two concepts?).

The key terms to think about here are: Concepts (what are we talking about), relationships (between concepts; customers place orders and so forth) and "things" (Product XYZ, Customer Thomas etc.). This is the realm of the business, not of IT. This is – essentially – what large parts of this book is about. There is no other way of achieving business governance on the conceptual level than to spend some time understanding the business of which you are a part.

2.1 Understanding the Business Using Concept Maps

Let us look at a very simple example, The EU-Rent Car Rental[1]. We will be following EU-rent throughout this book. To begin with here are some of the most basic concepts of a car rental business:

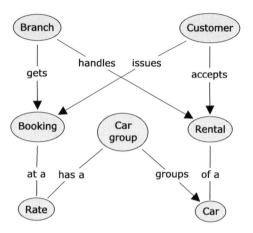

Fig. 2.1 High-level concept map of a car rental business ("EU-Rent")

Digression: The diagram is a "concept map",[2] which we will look at in detail in Chap. 6 below. The diagram really speaks for itself, which is the whole idea. Notice that you can read little sentences like "Customer issues Booking" and "Car Group has a Rate" and so forth. The sentences are visual – they are the connecting lines between the concepts. Concept maps can be drawn easily in brainstorming style workshops and they are easy to maintain, also for non-technical people.

Experiences from client projects (mainly in business intelligence contexts) since 2005 have been extremely rewarding. We are able to work with – on the business side – the concepts, their definitions and their relationships in intuitively understandable manners. Add speed, expressiveness and ease of use to the equation, and you understand why this is successful. Describing the business is a task for the business – use the right approach, keep things simple and control the business definitions together with the business people!

---

[1] Based on the accompanying example to the OMG SBVR standard (www.omg.org/spec/SBVR/1.0/). SBVR stands for "Semantics of Business Vocabulary and Business Rules". ACKNOWLEDGEMENTS: The EU-Rent case study was developed by Model Systems, Ltd., along with several other organizations, and has been used by many organizations.

[2] The technique of concept mapping was developed by Joseph D. Novak and his research team at Cornell University in the 1970s. See Chap. 6 for further details.

In essence, what you are working with is a set of structured terms (concepts) and their associated business rules (more on business rules later). Some things may change (e.g. sales channels) but the core of the business model remains conceptually much the same.

2.2 Information, Data and Business Rules

Many people are confused about the differences between information, data and business rules. This is a big subject. What is important here is the business perspective.

So what is the data to the business? Not a whole lot really. If we define data as those data, which make up your databases beneath your day-to-day operational systems, data is – at best – one of your sources of information. Mostly the databases contain detailed information about low-level stuff like item number, invoice date, quantity, price and discount. All are very appropriate, but not terribly exciting. First of all – you need more information than that to run a business. You also own so-called "unstructured data" – all your documents, emails, websites etc. And there are also external sources of data, which you need. Application data is just one source out of many.

Why is it that the low-level data is poor on information content, you may ask? Information is basically the answer to a question. That is why the invoice lines by themselves must be processed ("interrogated" is probably a more meaningful term) to provide answers to the key questions:
- Who, What, Where and When?
- And How Much?

This is called "Business Intelligence" (BI) – and most often you have to build a data warehouse in order to get all the information together in formats, which allow you to combine the data to give you the information, you need. In fact your ultimate goal is to answer the Why-question. That is where knowledge enters the picture.

The "interrogation" perspective is one way of describing the difference between data and information. Another way – perspective, if you will – is the issue of "meaning". According to information theory, data is just symbols, representing observations of ("pointers to") events happening in the world we live in. And the actors and things involved, of course. Information, on the other hand, is a message, which, if you share the context (common ground, if you will), will provide you with meaningful – well, here we are – *information*.

Compiling meaningful information (from data you have access to) is the basic process, which gives you knowledge.

The defining term here is really "meaning". Meaning is what enables you to gain information. If you think about this, it is a bit like a Black Swan[3] challenge. You should know what you know. You should also know what you want to know, but don't know already. And so on. How do you express "what I want to know"? *Meaning = concepts and structure.*

2.3 Going into Details: Business Rules

So much for "Information" versus "Data". Let us turn to business rules. Obviously, the simple "EU-Rent Car Rental" concept map diagram introduced earlier in this chapter is very high-level indeed. In real life you end up having multiple diagrams. But you also end up having a collection of other things – business rules. Here are some examples of business rules (from the car rental business context):
1. Each rental has exactly one requested car group.
2. The duration of each rental is at most 90 days.
3. Each driver of a rental must be a qualified driver.
4. If the drop-off location of a rental is not the EU-Rent site of the return branch of the rental then the rental incurs a location penalty charge.

If you think about this, some business rules could also be drawn up as concept maps. (In fact most can). However, some of the rules are quite detailed, and some also include "live data" (e.g. "90 days", or "SUV" etc.). You should use concept maps to get the overview and the structure in place. And then you supplement with the most important (and lasting) business rules (in plain text), which relate back to the concepts described in the maps. We use Excel or Word or similar office tools to record the business rules. See Chap. 6 for an example.

2.4 Good Definitions are Important

Describing business rules necessitates clear and precise definitions. Business concepts and strict definitions go hand in hand. In many cases you can do well with a pragmatic approach to this. However, there may be reasons (complexity and reliability for instance), which necessitate a strict and systematic approach to this.

Definitions of core business concepts last a long time and reach many people. This means that they should not only be precise and to the point. They should also communicate well and be easy to remember. In many respects production of good definitions is a skill that requires you to be conscious about, what you are doing. Fortunately this skill is something that can be refined and rehearsed. There is no better place to look for expertise about good definitions than this excellent book: "Definitions in Information Management" (Chisholm 2010).

---

[3] Nassim Taleb, "The Black Swan: The Impact of the Highly Improbable", Random House, 2007.

When we do concept mapping – as described in Chap. 6 – we do not care too much about detailed logic in the diagram. Concept maps are not about logic, they describe concepts and their relationships. In business rules, on the other hand, you typically have a lot of logic. Logic is very necessary when you specify *something you want to include in an IT-system* of some kind. However, on the business level, the world is often a bit fuzzier and we really do not need that extra precision, which logic adds to the picture. Simple business rules formulated in clear language is more than enough in most cases. From a business perspective what is stable is concepts and relationships. They survive most things, including change of ERP-systems, for example.

And people are not 100 % rational. A good example is the (systematic) errors most people make when reasoning with probabilities. Add intuition, gut feelings and so forth; this complex set of capabilities is what we are and this is what we should support.

2.5 What Do Concepts Actually Look Like?

Let us focus on concepts and their relationships.

The EU-Rent Car Rental fictitious business, which was mentioned earlier, is described by in the OMG standards documentation as follows:

> EU-Rent rents cars to its customers. Customers may be individuals or companies. Different models of car are offered, organized into groups. All cars in a group are charged at the same rate. A car may be rented by a booking made in advance or by a 'walk-in' customer on the day of rental. A rental booking specifies the car group required, the start and end dates/times of the rental and the EU-Rent branch from which the rental is to start. Optionally, the reservation may specify a one-way rental (in which the car is returned to a branch different from the pick-up branch) and may request a specific car model within the required group.
>
> EU-Rent has a loyalty club. Customers who join accumulate points that they can use to pay for rentals. EU-Rent from time to time offers discounts and free upgrades, subject to conditions.
>
> EU-Rent records 'bad experiences' with customers (such as unauthorized late return from rental, or damage to car during rental) and may refuse subsequent rental reservations from such customers.

The text above is pretty clear and well written. However, experience has shown that concept maps are indeed much more intuitive to read and understand. The diagram below represents almost all of the information in the text above:

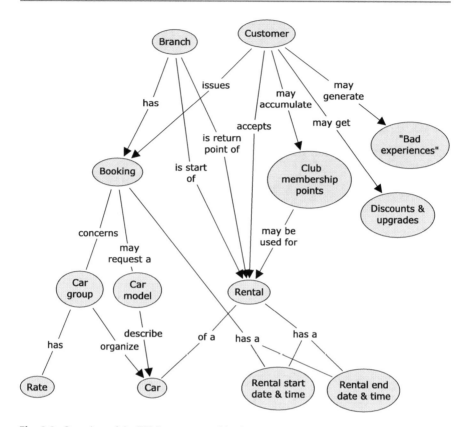

Fig. 2.2 Overview of the EU-Rent car rental business

In addition to the concept maps, we need two "documents":
1. The verbal definitions of the concepts (cf. Chap. 6 for details), and
2. The necessary – longer term – business rules (written in clear language).

Verbal descriptions could be along these lines: "Concept: Car Movement. Definition: Planned movement of a rental car of a specified car group from a sending branch to a receiving branch".

This is no big deal. Many organizations handle this quite well in Microsoft Word or Excel.

This concludes the introduction to understanding what the business means. In the diagram (concept map) below, you can see an overview of the role of business understanding in the whole business development initiative:

2.5 What Do Concepts Actually Look Like?

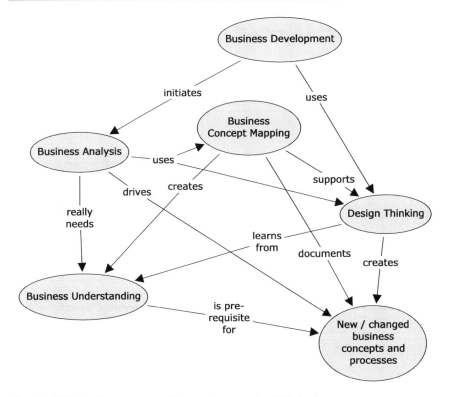

Fig. 2.3 High-level concept map of the major proposals of this book

Business understanding is created via concept mapping, which we come back to in subsequent chapters. From the business analysis point of view the reason to create concept maps is that design thinking people learn from and are inspired by the business understanding – expressed as concept maps. In that way concept mapping is a prerequisite for the creation of new/changed business concepts (and processes). Concept maps trigger the "A-Ha!" moments, which design thinking people get excited from. Since the most important issue is to get that design thinking going, we need to understand in more detail what the design thinking paradigm is about.

Design Thinking for Business Analysis

3.1 Business Analysis: Understanding Business Information

If you think that Business Analysis is all about SWOTs and RASCIs, you should prepare for some re-architecting. All those tried and tested tools and methods are still there, of course. But they are missing the most important point: The business information!

Much effort has been spent over the years on analysis and reengineering of business processes. And that is very good. The nice thing about business processes is that you can always change them, again. For instance as new information and communication technologies become available.

What is much more stable and unchangeable, though, is the business information asset. Imagine you are in the real estate business. There you have some important concepts – like properties with rooms and windows and so forth. And the most important thing about properties is – you guessed it – location! Those concepts are not going to change any day soon.

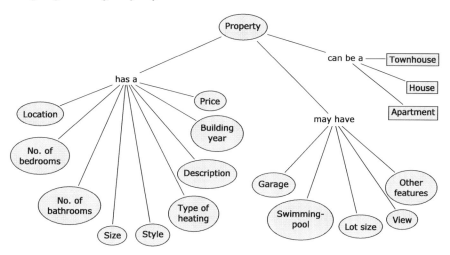

Fig. 3.1 Real estate property concepts

Most people acknowledge that the key concepts of the business are important. However, they assume that they are all readily available as well-defined stuff, which everybody agrees on.

Not so. Picture the enterprise as a very large building with many floors, endless corridors and lots of doors and windows.

Fig. 3.2 An "Introvert" office building in Copenhagen, Denmark (Photograph by the author)

You may have some ideas about what is behind most of those windows and doors. But you cannot know in detail. And you are bound to be wrong on some points. If you do not need to open the doors, you do not have a problem. Or?

Sooner or later you will have to "open a door" and see what is behind it. Why? Because you must – or would like to – change something. Change – hopefully to the better – is a very important driver. As you open the doors, one at time, you have to be able to put your findings into the context of what you already know.

This is very similar to the situation a designer is in. Let us see how we can deal with that in an intelligent manner.

3.2 Design Thinking: Where Does It Come from?

In an interview in Academy of Management Learning and Education Roger Martin, dean of the Rotman School of Management in Toronto, describes design thinking as having three aspects: cognitive, affective, and interpersonal (Martin 2006). The affective and interpersonal skills are not in scope of this book, but cognition is well inside the scope. The cognitive aspects are the three types of reasoning: inductive, deductive, and abductive.

In classic logic, inductive reasoning is generalization from specific instances, while deductive reasoning involves inference from logical premises.

MBA programs provide students with both inductive and deductive reasoning, but under-emphasize abductive reasoning. Abductive logic is the process of forming an explanatory hypothesis (Martin 2009). A classic example of abductive inference is:
1. It rained last night.
2. The lawn is wet.

The fact (2) that the lawn is wet makes the hypothesis (1) quite plausible. Other hypotheses exist, but are (depending on circumstances) less likely. So, by abduction we conclude that it must have rained last night. Of course, in real life you will have more than one possible hypothesis. The trick is to somehow identify the most plausible of them. This is often related to simplicity, elegance and/or some kind of economics. The best explanation is the one that makes most sense. To give a general example: We observe a surprising fact, C. Now, if A (something) were true, C would be a matter of course. Hence there is reason to suspect that A is true. At least A *ought* to be true.

As you can see, abduction is a very natural sense, really. It helps us navigate the world by filtering in only necessary information. Closely related to track hunting, maybe.

3.3 Designing Other Things than Products

Tim Brown, CEO and President of IDEO (a leading design firm) writes in Harvard Business Review: "... as economies in the developed world shift from industrial manufacturing to knowledge work and service delivery, innovation's terrain is expanding" (Brown 2008).

In other words: From being restrained to physical products, design can and should be applied to processes, services, IT interactions and in general communication and cooperation. *One of the key purposes of this book is to add business information analysis to the list of design thinking approaches.*

Over the last few years more and more business people have accepted the idea of "Design Thinking" as an approach to business development. Not only of products but also of services, the organization, the business models, the narratives and all the rest of business management should be *designed*, not *engineered*.

In a report for the Finnish Ministry of Employment and Economy, Provoke Design Oy quotes A. G. Lafley, CEO of Proctor and Gamble, for this (Aminoff et al. 2010):

> Business schools tend to focus on inductive thinking (based on directly observable facts) and deductive thinking (logic and analysis, typically based on past evidence). Design schools emphasize abductive thinking - imagining what could be possible.

Proctor and Gamble was into design thinking early and established their Global Business Services (GBS) from that. GBS is, generally speaking, a project-oriented culture. This – among other things – helped integrate Gillette quite smoothly. See also (Martin 2009) about this case story.

3.4 Dealing with Wicked Problems

Within the design thinking movement there is a lot of talk about "Wicked Problems" – problems that are ill-defined (Martin 2009). In our context here is this book, the both problem and the solution might be unknown at the beginning of the business development project.

Let me – based on extensive experience with the business analysis parts of business intelligence projects for different clients – describe a typical project situation at startup time:

- We know that we want to support the sales and marketing people with business intelligence based on data in an upcoming data warehouse.
- We do not have a model of the business information and we do not have meaningful definitions of the concepts involved.
- We do not know what data we really have in the databases.
- We do not know which potential problems there are in the data (like eg. missing information, which we thought was there).
- We do not know in detail how the source systems are being used by the end-users.
- There are conflicting messages about what information is needed to support the daily operations across the different departments and divisions.
- We do not know which key performance indicators we want (not to say why we want them).
- We do not know what kind of budget we need to have for doing this or how long time it is going to take.

Here are the same unknowns in a slightly more systematic way;

Typical data warehouse project start-up conditions	We know for a fact	We do not know
Aware	We want to build a data warehouse	Business concept model and definitions
	Differences of opinion exist	What data do we have?
		Which key performance indicators do we need?
		Which budget and how long time is necessary?
Un-aware		Incomplete or missing data?
		How is the source system being used?
		Why is it we need those key performance indicators?

Is that a wicked problem? Most will agree.

Obviously, business intelligence projects are not normative for everything else, but they are very typical business development projects. (Or they should be, rather).

The consequence is that a large part of the project is spent on understanding, defining and creating the design of the solution. Now – that is more akin to what an architect does than to what an engineer or an economist does. And the architect uses design thinking, obviously.

3.5 Design Thinking for Business Development

This book is not about design thinking per se. However, much of what is going on, when you discover and manage the business information asset using concept maps and definitions, is really design thinking. Here follows only a brief overview of design thinking in general (in the business development context). Following that we will have a closer look at how concept mapping and design thinking fit together.

Let us start with the class of social system problems called Wicked Problems, which we described above. Roger Martin (Martin 2009) says:

> Whereas managers avoid working on wicked problems because their source of status comes from elsewhere, designers embrace these problems as a challenge.

The dividing line is not between synthesis and analysis, but between "reliability" and "validity".

Reliability is the traditional business management approach. Managers are rewarded for providing measurable results (shareholder value and so on). Investors and shareholders and bonus programs use analytical approaches to determine the reliability of the business. This is the world of charts of accounts, budgets, forecasts, corporate performance management, business intelligence and – not least – quarterly results.

On the other side of the house, designers are rewarded for **validity**. Producing things that really are valid (on the market, in the organization, among the customers, employees etc.). It is the validity angle, which is in focus in design thinking. Validity and innovation go together. Validity is meant in the business sense. If a change or a new thing provides business value (in new ways) it is valid, otherwise it is not. Validity comes by design. Reliability comes from "engineering" – robust systems, procedures, repeatability and so forth.

This difference between the traditional business analysis methodology and design thinking approaches is also confirmed in "Designing for Growth" (Liedtka and Ogilvie 2011). The traditional business method is analysis aimed at probing one "best" answer whereas designers experiment with the aim of iterating toward a "better" answer. Creating the new while preserving the best of the present.

Roger Martin depicts the way business changes happen as a "knowledge funnel" (Martin 2009). In the top you have "mysteries" – the wicked problems, all those little doors you can open only to discover that you do not understand what you see. The trick then is a two-step series of events:
1. Turn mysteries into heuristics (rules of thumb); this enables you to get ideas and to prototype solutions, which you can test, improve, test again and so forth.
2. Turn heuristics into algorithms (precisely defined programs and procedures), which can run your business and generate that reliability, which the stakeholders are so interested in.

What you have to do is to "make an inference to an explanation". And the explanation should be "the best explanation one can devise given the data, which is insufficient to yield a statistically significant finding. (Abductive reasoning)".

The findings (the explanation) "produce a prototype and observing whether it operates as desired or expected" (Martin 2009).

The Danish poet/artist Martin Bigum has reworked the concept of "synchronicity", which originally was coined by C. G. Jung in the 1920's (Jung 1993). In Martin Bigum's reformulation, he defines the phenomenon like this:

"Synchronicity can be explained as an experience, which at times arise in the human brain, as it encounters the disparate events of reality and interprets them as coherent or meaningful. Things and events, which appeared to be random when they occurred, turn up to be part of a greater picture - a structure beneath the skin of the daily life". (In a Danish language exhibition catalogue, "Structure beneath the skin", 2012). This way of saying it is a beautiful description of the "design thinking secret".

3.6 Business Synthesis

The focus of design thinking is on synthesis, not analysis. So we need to become "Business Synthesists". Concept mapping is the preferred tool because synthesis is much about writing narratives. Concept maps are great storytellers! They are designed for it.

Something to think about: *Even the chart of accounts can be thought of as a kind of narrative* (Martin, 2006) involving concepts and relationships. This will be described in Chap. 5.

The process of developing concept maps in brainstorming sessions is highly supportive of creating understanding of different angles and perspectives. The A-Has are bound to appear sooner than later.

This is the cheapest way to get real business pains out in the open! Business pains survive by staying in the dark! Besides clarity (avoiding misunderstandings) and having everybody "speaking the same language", increased focus on business concepts promote a better and more detailed understanding of real business pains.

Please note that using design thinking does not imply that you design your as-is situation! Using concept mapping you describe the real world of your business as it is. This is why we call it concept *mapping*, not concept design. This process (the discovery and the subsequent understanding) is one of the skills designers have. You describe and understand – and get ideas on the fly! Then you improve and apply. At this point *designing* a new solution is implied, possibly also new concepts or generalizations of such, what ever it takes to do the job.

From the concept mapping perspective, it is the first two parts of the transformation process of design thinking, which are really interesting.

Tim Brown puts it this way:

"... the ability to see all of the salient - and sometimes contra-dictionary - aspects of a confounding problem and create novel solutions that go beyond and dramatically improve on existing alternatives" (Brown 2008). Two of the necessary skill sets are experimentalism and collaboration.

Concept maps do all of that. And they are easy to produce; they immediately improve the understanding. They can be easily redrawn, changed, combined, tested or thrown away if another approach is found to be the best explanation.

Liedtka and Ogilvie mention ten tools for doing design work. The first and most important tool is "a visualization" tool: "This is really a meta tool so fundamental to the way designers work that it shows up in virtually every stage in the process of designing for growth". Clear communication is of essence. Presenting ideas as drawings makes them easier to understand and reduces the risk of misunderstandings (which are plentiful if you only describe things in text). Before the ideation phase it is necessary to get a deep insight into the problem / opportunity and its context. They recommend brainstorming in the ideation phase followed by the use of a concept development tool. In concept development, visualization is again of essence. "Concepts are literally coming out of your imagination and your brain is creating pictures of something that doesn't yet exist" (Liedtka and Ogilvie 2011).

3.7 Where Does Concept Mapping Come from?

Using design thinking to enable business development and change requires understanding where you are today. There are reasons why concept mapping is the perfect match for this (Novak 1990, 2008), (Novak and Cañas 2006), (Moon et al. 2011).

Concept maps were originally designed and developed by Prof. Joseph Novak (1990, 2008) with a learning focus. The theory behind concept maps is based on D. Ausubel's work on meaningful learning – see (Novak 2008) for a good overview. Assimilation theory (as it is also called) suggests that learning is based on representational and combinatorial processes, which occur when you receive information. New concepts are related to relevant, existing concepts in an existing cognitive structure (in a non-verbal representation).

In other words, there are two processes: Discovery (of information) that leads to reception of the information – integrated with what the learner already knows. In this manner concept mapping is not only facilitating learning, but also creativity. Notice the parallels to the Inspiration-Ideation-Implementation in Tim Brown's view (Brown 2008) as well as to Roger Martin's flow from Mysteries to Heuristics to Algorithms (Martin 2009).

The key is to depict the relationships between the basic concepts and terms that will be used. Learning then takes place by assimilating new concepts into existing conceptual structures known to the learner. That is precisely what Prof. Novak designed concept maps to do (Novak 2008). And that is why they have proven to be well suited for the inspiration and ideation phases of the design thinking approach. By way of concept mapping (the diagrams), we make knowledge explicit.

Another researcher, Rivka Oxman, puts it this way: "Conceptual knowledge, the ideational basis of design, constitutes one of the most significant forms of knowledge in design. Concepts are fundamental to design thinking, since they operate on an ideational level. They are the fundamental material of design thinking" (Oxman 2004).

Drawing concept maps is a learning exercise. So is reading and discussing them. Changing them together in brainstorm sessions creates a very creative environment. And add to this a lot of new business understanding as well as new opportunities.

3.8 When Do We Need Tools Like Concept Mapping?

You use concept maps in many different situations, for example:
- To give a high-level conceptual overview (setting the scope, maybe) in the beginning of the project.
- To explore the "mysteries" you often develop a set of concept maps, which are re-iterated over some time. They can describe as-is and also "wannabe future" concepts.
- Prototype solutions are also mapped conceptually until you arrive at a working prototype, which you decide to go along with.
- The final solution must have a conceptual design, of course, and you may also add various detailed documentational or instructional concept maps to explain things to the business people.

What you need to do is to open all necessary doors and look behind them. Remember that some doors are hidden, secret or forgotten... Once you know what you have, you map it – producing "floor plans". Having those in place you need to figure out where the hot spots are and how you can make it as easy as possible to get from one of the "good spots" to the other.

Tim Brown calls the first phase "Inspiration" (Brown 2008). In our context we prefer to call it "Exploration" although the main activities are similar. Inspiration is too narrow. There really is a substantial amount of exploration going on. Actually, you might also say that the first time is spent on learning. Even old-timers can learn new things about what is actually going on. It is this explorative learning process that creates the moments of inspiration and insight, including "A-Has".

For the following two groups of activities, we lean on the terminology of Tim Brown. This gives us the following flow:
- *Exploration*, which among other things contains the organization of information (concepts and relationships), the sharing of insights, storytelling, and synthesis of possibilities (more stories). Concept maps are fabulous storytellers.
- *Ideation*, which among other things involves making sketches, scenarios, more storytelling and internal communication. Again, concept maps excel in all of this.
- *Implementation*, where concept maps are very good for documentation, communication and instruction.

There is a flow of conceptual structures going through the business analysis projects:

3.8 When Do We Need Tools Like Concept Mapping?   23

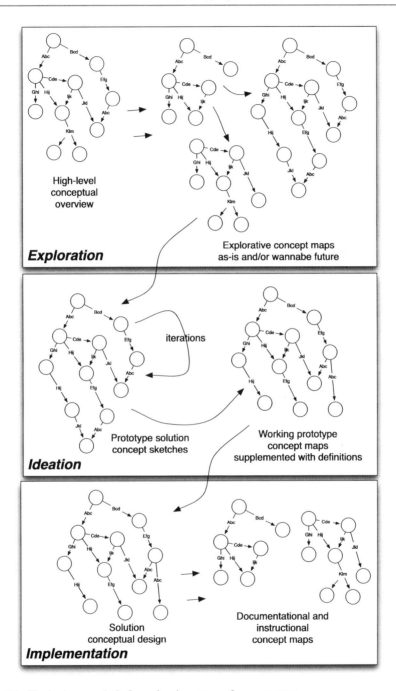

Fig. 3.3 The business analysis flow of various types of concept maps

With both concept mapping and design thinking having been described above, it is now time to look at how the combination of the two actually change the way the business analyst works.

Business Analysis Redefined 4

4.1 Overview of the Method

As you have seen, the design thinking approach consists of three stages and of different sub-categories of activities, when it comes to business analysis:

Stage	Activities
Exploration	Sharing of insights
	Storytelling
	Organization of information
	Synthesis of possibilities
Ideation	Sketches
	Scenarios
	More storytelling
	Internal communication
Implementation (of a design)	Making design decisions
	Documentation
	Communication
	Instruction

Parts of the above are based on (Brown 2008).

We are interested in business information, not very much in processes. Needless to say, business process management is important, but we are nevertheless not going to look deeply into the processes, only the information side, because information last longer.

Where then is the knowledge about the business? Obviously there are a lot of potential sources:
1. Inside people's heads
2. In the business model documents, strategy documents, business plans, instructions, administrative guidelines and hundreds of other documents
3. Inside day-to-day documents such as charts of accounts, spreadsheets, IT systems, engineering documentation, marketing materials, websites, specifications of

requirements and of systems, IT data models and databases, business intelligence reports and their "multidimensional cubes"

Option (1) above is by far the most reliable and the easiest (and most productive) source to work with. In other words, what is needed is a series of "debriefing" workshops followed by ideation and synthesis workshops.

Working directly with the business people in the analysis and synthesis stages also creates opportunities to:

- Create new awareness and "A-Ha moments".
- Gain real knowledge about what is actually going on.
- Get the real pains out in the daylight.
- Rethink terminology.
- Discover conceptual "holes", contradictions and mistakes.
- Be creative – get new ideas.
- Play with scenarios.
- Get acceptance of the structure, the terminology, the definitions and the business rules.

Design thinking approaches, brainstorming, quick sketches, visual storytelling and dialogue go intimately together. That is why concept mapping is such a powerful facilitator of the whole flow.

On the other hand the analyst will also have to go to other sources such as the ones mentioned above. We will get back to that in Chap. 5.

4.2 Preparing for the Analysis to Synthesis to Design Flow of Events

Obviously you will have to prepare yourself with an up-front understanding. A very productive way of accomplishing this is to do some concept mapping for yourself as you go along reading and inspecting various sources of business information content.

Analysis of a business area and subsequent synthesis of a changed or new design of business (Martin 2009) is of very high importance to the business. You cannot do it without participation of the business people. Even the top people of the business or business unit should be involved. Depending on the sponsor of the project consider having e.g. the CFO or the Director of Sales or similar kinds of executives present in the start of the project as well as in the final acceptance review meeting. For all of the workshops you need a "project anchor" in the business organization, could be a senior controller maybe. He or she should be on the project from start to end. In addition to that you need 1 or 2 really knowledgeable business specialists – the kind of people who really know what is going on. They are typically in line management staff functions. And then you need – at times – some real business people present, be it sales people, logistics et cetera as appropriate.

What you are going to go through – together with important business people – is a very intense and interactive process stressing your communication skills and your empathy. Be a very good listener and be a very good questioner. Use open questions and open language. Use metaphors to open the minds of the business people. See for example "Clean Language" (Sullivan and Rees 2008) for good advice. People skills are also of essence, together with a lot of patience and polite exploration of the context people are defining their experiences from.

What then characterize people, who master design thinking? Three traits are essential: Intuition, synthesis and originality (Martin 2009). Intuition can (most likely) be refined; look at the difference between an experienced track hunter and a city dweller, when it comes to reading signs on the ground in the forest. Synthesis is somehow related to spatial cognition (Gärdenfors 2000, 2004). Much of synthesis is about causal relationships (Martin 2009) and here concept maps really help improving the cognitive processes. Some people have better spatial capabilities, but it is likely that this can also be trained and refined. Originality is related to risk-taking. In other words: willingness to experiment and change (Martin 2009). Experience certainly helps.

Remember that the starting point is validity driven thinking. People, who are capable of doing that, are often the kind of people, who are brought in to get the organization out of the box (Martin 2009). Validity is not better than reliability or vice-versa. Both are very fine qualities in their own right. Knowing the difference between them is so essential to thinking clearly about what is business and what is "engineering". If you end up mixing the two, you just get exactly the current state of affairs in analysis and design. And we want to do better than that! On the business concept level **validity** is the quality to strive for.

Be as ready as you can before the first workshop.

4.3 Top Down: The First Workshop

The first thing to accomplish is to describe a common ground for the project on a high, overall level. In other words you must develop a one-page overview concept map together with the business people. You may prepare a rough sketch in advance based on high-level business documents. Only high-level business objects (possibly generalized like "Production Line" in a multi process manufacturing flow) should be included.

What must be done is to find a way to start "opening the doors" in an efficient manner. The business model formulates some of the most essential facts about the nature of the business and could be a good place to start looking. The exciting book "Business Model Generation" (Osterwalder and Pigneur 2010) identifies the nine building blocks of business models: customer segments, value propositions, channels, customer relationships, revenue streams, key resources, key activities, key partnerships and cost structure. Expressed as a diagram (they call it the "Business Model Canvas") it looks like this:

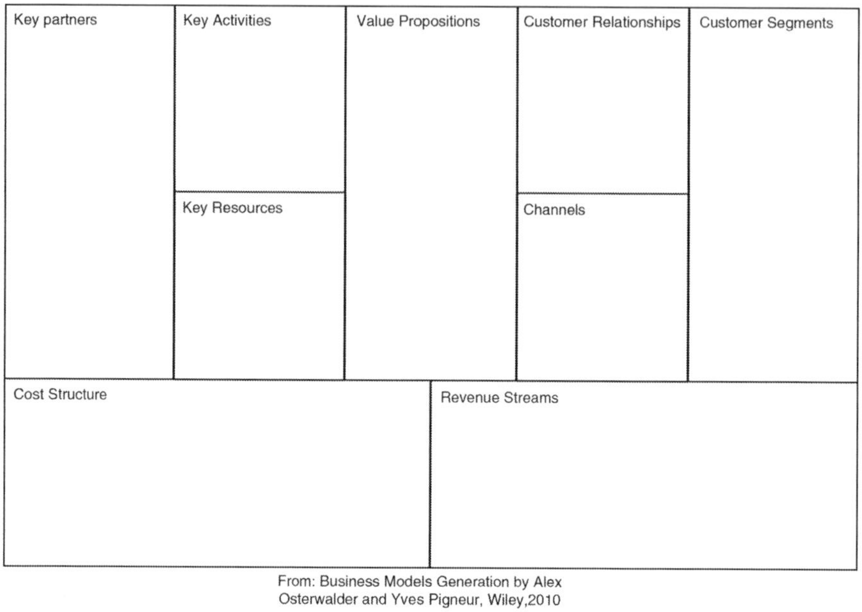

4 Business Analysis Redefined

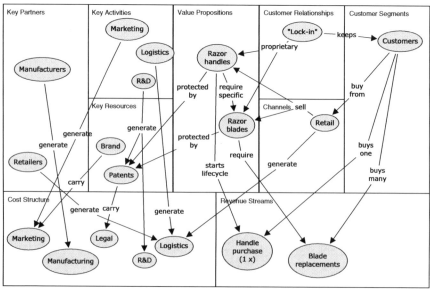

Fig. 4.1 The business model canvas (Osterwalder and Pigneur 2010)

Actually the combination of the business model canvas and concept mapping works very well. This is the very well known business model of Gilette (Osterwalder and Pigneur 2010) layered on top of the business model canvas:

Fig. 4.2 The Gilette business model on top of the business model canvas (Osterwalder and Pigneur 2010)

4.3 Top Down: The First Workshop

Take a moment to read the sentences of the concept map. And to understand why the concepts are placed where they are on the business model canvas. This could be a good starting point for a discussion on the first workshop, where representatives of the management are present.

Obviously there are many other ways to establish this, the first "common ground", concept map. Here is a simple example from a car dealership:

Fig. 4.3 Early, overview-level, concept map of a car dealership

You should be able to produce similar overall concept maps in a couple of hours – brainstorming-wise with live concept mapping as you go using e.g. CmapTools and a data projector.

Having reached this point, you are ready to scope the project. Very likely ambitions have been too high and you will have prioritize and split the subject area into a couple of subsequent projects.

Get a good understanding from the representatives from the upper management about their concerns, their goals and ask them directly what keeps them awake at night?

If you accomplish this in 3–5 h, you have done well. In general it is recommended to keep brainstorming sessions not longer than 3–4 h. Most people loose the intense concentration, which is required, by then.

4.4 Explorative Workshops

The next 2–4 workshops should be focusing on:
- Organization of information (concepts and relationships)
- Sharing of insights
- Storytelling
- Synthesis of possibilities

The first three parts are efficiently facilitated by (live) concept mapping.

You will have to break down the subject area into manageable parts (corresponding to one-page concept maps). Each of those you break down in some detail:
- The actual day-to-day business objects
- Selected, key, properties, including measures
- Possibly some exceptional business rules

Do not attempt to describe everything at this stage. Focus on the essence of the business information, not the details (they will come later).

What we do in the exploration stage is this:

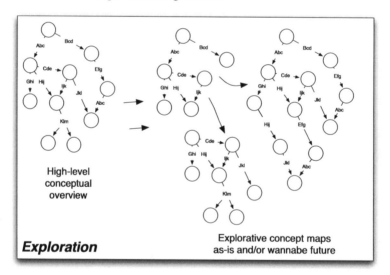

Fig. 4.4 Exploration stage concept map types

We explore – using concept maps to share insights, to tell the stories and to organize the information. But what is it we explore? Obviously it is – to a large extent – the business concepts and relationships as they exist today. The as-is situation.

However, since some business development driver initiates most projects, we actually – at this stage – begin to explore possible "wannabe" future concepts and relationships.

Already at this stage we get some surprises. And surprises drive the design thinking approach. Here is a selection of possible discoveries:
- We lack information about something, which could be of importance.
- Our terminology is imprecise or arcane, born out of tradition maybe.
- Some concepts are plain wrong.
- Is this really what we are doing? (And other "A-Ha" moments).
- We don't have information about this.

At the end of this stage we know that there are pieces missing, that there is conceptual rework to be done and that there is room for improvement in terminology and IT-system support, possibly.

In particular the A-Ha moments are interesting and should be carefully remembered. They are certainly idea generators later on.

After 3–4 h you have produced 1–3 concept maps (in rough form, but definitely sense-making). Chances are you will need more than one workshop. Now you have "the floor plans" of the enterprise expressed as concept maps.

4.5 Ideation Workshops

The ideation stage is where changes are being created (designed). The output from the exploration stage is a set of explorative concept maps. The beginning of the ideation stage is, consequently, set in a review context. One at a time, we look the first concept maps over. Then we get to design work. There are both small and larger tasks to do. The work is again organized along design thinking lines:
- Sketches
- Scenarios
- Storytelling
- Internal communication

This is the concept mapping, which goes on in the ideation phase:

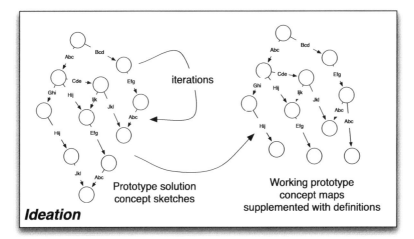

Fig. 4.5 Ideation stage types of concept maps

For all of this, concept maps are the core vehicle for driving the processes. Here is an example from a client case (public sector) of a sketch of something in search of a design:

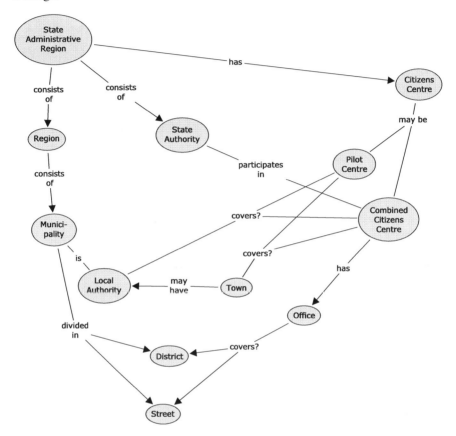

Fig. 4.6 Public sector example of a prototype concept map

Remember that we – from the exploration stage – have:
- Holes to fill (missing concepts and relationships)
- Concept "repair and restoration" work
- Doing things better than before
- Learning from the A-Has
- New design of new things

There are a few other techniques, which are useful in the ideation stage (Liedtka and Ogilvie 2011):
- Explore extremes
- Change the actors
- New scenarios and new trends
- Contradictions
- What if we were someone else?

4.5 Ideation Workshops

– Try to stand in the future and look back
– New combinations

Another important way of creating ideas is to prototype. Prototyping may happen as concept maps – testing the validity of different approaches. But prototyping can certainly also be data based prototyping using simple tools (like e.g. Microsoft's PowerPivot, Tableau from Tableau Software and many others). Looking at data can – at times – reveal "hidden secrets" and other versions of the reality than the one subscribed to by most people in the organization. The combination of concept prototypes in concept maps and quick "proofs of concept" based on live data is very powerful.

This phase, the ideation phase, is the most important process of the three. It is here that business value is created as you exploit what you learned about the enterprise in the exploration phase. The design thinking approach helps the business people and the analyst to transform the reality of the business and lays the foundation for building new and better conceptual solutions. If you feel that you are unable to force creativity to the surface, it could well be that your level of understanding is too small and maybe incomplete. You will need to go back and explore some more in order to see whether you can find some places, where "turning things upside down", for example, can help you be creative in your subsequent design work.

Experience shows that by using live concept mapping, you will arrive at good solutions – but it might take some time. Some problems are hard ("wicked") and cracking the nut can be difficult. The more sketches, scenarios and storytelling you visualize, the faster the ideas arrive.

Likewise, working with the terminology frequently helps you improve the solution substantially. Here is an example from the logistics area:

A logistics company is in the business of packing advertising brochures and newspapers for delivery to distributors, who distribute them to consumers on carefully planned distribution routes. Traditionally they thought of the packing as consisting of:

- Machine packing (producing letterbox ready bundles of wrapped brochures etc. on machines)
- Manual packing
- Packing by distributors

In other words, their day was full of very low-level and disparate information arriving from quite different types of processes. During a couple of sessions we reworked the concepts a bit. The fact is that they are in the business of producing distribution packages. So everything is driven from that central concept. This really loosened up the information visibility and led to much better insight. It also meant that the value chain can be followed elegantly in the data warehouse. What the distribution department does is to distribute on routes – for which they receive distribution packages, which are delivered to them by the logistics department.

The ideation phase thus consists of a number of attempts to find alternative ways to improve some aspects of the business. To stay in the enterprise as a building metaphor: Can we break down a wall, maybe? Install a new staircase? Move some offices around? And so forth.

Here is another example of a concept map of a little business area having found some good solutions:

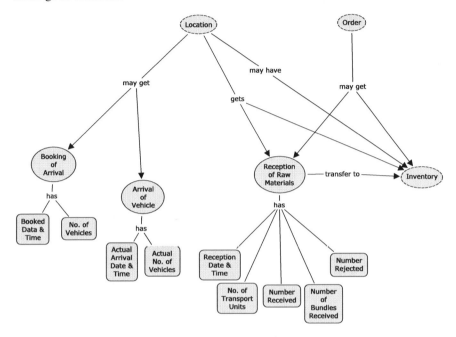

Fig. 4.7 Potential solution concept map from a logistics context

Note: In the concept map above dashed outlines indicate that the concept is described in detail on another diagram. This is often used because you may end up having hundreds of concepts – too much for one piece of paper.

4.6 Generalization and Specialization

One of the most powerful tools available to humans is what we call "generalization". The opposite, specialization, is also quite useful.

In conceptual analysis you – at times – get stuck. The reason for the halt may be too much complexity. In that situation generalization is maybe a way forward. You may be looking at 5–15 different concepts having complex relationships. Obviously you can simplify by introducing a new concept (probably previously unknown).

4.6 Generalization and Specialization

A complex, official address structure:

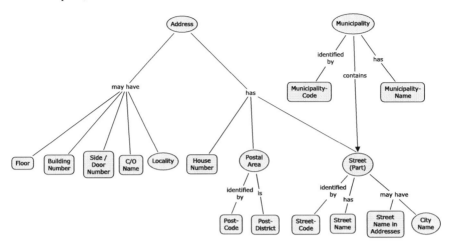

Fig. 4.8 Concept map of official address concepts

A simple address structure:

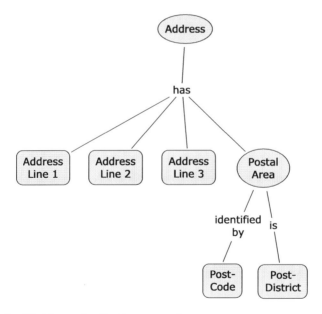

Fig. 4.9 A simplified (generalized) address concept map

The new, generalized concept may – by itself – be the solution to a "Gordian knot". You then have a choice: Is the detailed information (concepts and relationships) really necessary? If the business can live with the more general concept – that is fine. Get rid of the unnecessary complexity. Generalization is a very powerful technique for reducing the size of the scope. "Do not leave home without it!"

On the other hand, there are situations, where you are stuck due to lack of information. In that case you need to specialize, Get into the details and take it from there. Obviously this only works when having a dialogue with the business people. Who else would know how to "go down one level"?

4.7 Levels of Abstraction

Often there is some confusion as to which level of abstraction you are looking at. For example: Do we (in this organization) talk about "Cars" or of "Vehicles" or both? Is the concept of "Engine Size" a property of Cars or of Vehicles? And so forth. Getting this wrong can lead to quite expensive reworks of data models and applications in the IT systems down the road.

4.8 Implementation Workshops

Remember that the ideation stage concludes on some possibly alternative solution prototype concept maps. Business managers should come on stage again for prioritizing the solution alternatives and make the final decisions about which they prefer.

The rest of the final stage is about designing the solution in detail. This involves designing final, detailed concept maps, which document and communicate the concepts and relationships to be used onwards in the business.

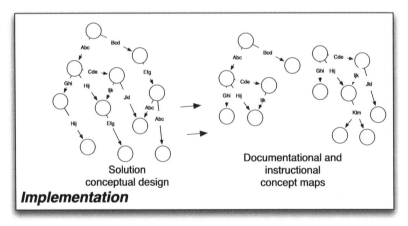

Fig. 4.10 Implementation stage types of concept maps

The concept maps live on as parts of system documentation, administrative guidelines, intranet information, course materials for new employees and so forth.

4.8 Implementation Workshops

Here is an example of a solution design on the concept level:

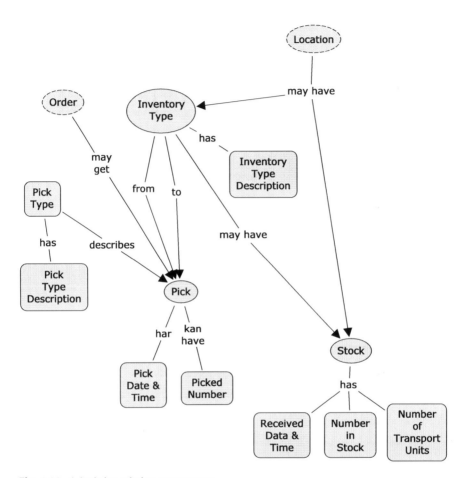

Fig. 4.11 A logistics solution concept map

Depending on whether the solution should be supported by information technology, the design can also take shape of e.g. a logical data model. Here is an example of a way of representing a multidimensional structure (a star schema, really) in a business intelligence environment:

Fig. 4.12 A business intelligence solution concept map

A client has nicknamed this particular style of diagramming a "peacock"! Within business intelligence modeling the structure this way is a new style of representing what is called a Star Schema in the BI jargon.

In this phase what we design is the input to the actual construction phase. Having done that, we can let loose the masons and carpenters and start reshaping the future.

4.9　Agile Approach

The business analyst can move on to the next subject area, and perform the same process there. Wall-to-wall business concept mapping in a midsize company is certainly doable. Obviously you will do some up front business priorities and only initiate the higher-priority activities – from a business priority point of view.

However, a more agile approach is also viable. Once you have a good, top-level overview of the key conceptual subject areas (using a few overall concept maps), you may set up a series of sprints, where you do exploration, ideation and implementation on subdivisions of a subject area. Limit yourself to one-page concept maps, and you can progress quite rapidly. It is recommended, however, to have at least two sessions (workshops) to give the sub consciousness some time to work with the conceptual structure in tranquility. At times it is amazing to see that as you pick up a concept map the second time, you suddenly get insights, inspiration and those exquisite "Oh – now I see it" moments!

4.9 Agile Approach

A typical flow of event in an ideation stage could look like this:

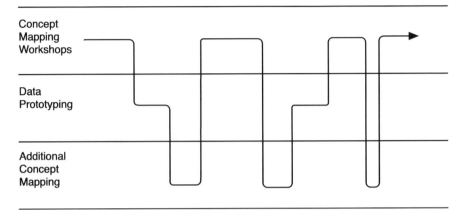

Fig. 4.13 A example of a flow in an ideation stage

Here is (again) the concept map of the first two sections of the book:

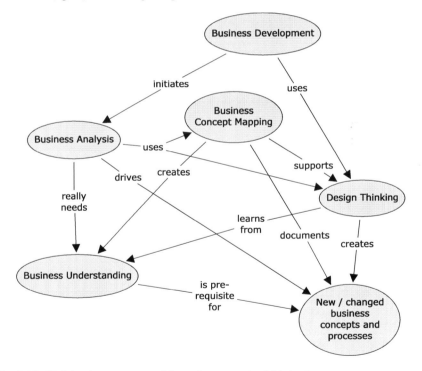

Fig. 4.14 High-level concept map of the major proposals of this book

The next step is to get the business concept mapping started – where to go looking for business concepts and how to map them.

Part II
Business Concept Mapping

"The beginning of wisdom is the definition of Terms" Traditionally attributed to Socrates (469 BC-399 BC)

5 Where to Find Meaningful Business Information?

Business information, in particular business concepts are, of course, everywhere you look. This chapter will provide a number of examples of what to look for and where to look. Start with the Business Model and proceed to e.g. the Excel sheets, the charts of accounts, databases and applications, reports, documents, intranet content, the internet and so forth. Even Excel-formulas may be represented as concept maps. This makes a lot of sense. Some of the most widely used Excel-sheets contain very important calculations of e.g. key performance indicators, which are essential to running the company. The business concepts and relationships between them are some of the organizations most important pieces of knowledge. You should get that knowledge out in the open.

You will realize that some of the knowledge about the concepts used in your particular business is found only inside the heads of a few persons. That is not as it should be; it is a (considerable) risk to your organization. This means that you will have to include brainstorming sessions in the process of "harvesting business concepts". That tacit knowledge inside the heads of a couple of a few key people can be quite essential. Those are the most important "doors" to open in order to explore what is in there. Afterwards you then need to include them on "the floor plan" – the concept maps of the business area.

The purpose of this chapter is to raise the awareness of what business people are looking at in their day-to-day activities as well as in reporting and analysis.

5.1 Start with Your Business Model

"Business Model Generation" (Osterwalder and Pigneur 2010) was introduced earlier in this book. It is based on nine building blocks of business models: customer segments, value propositions, channels, customer relationships, revenue streams, key resources, key activities, key partnerships and cost structure.

The Business Model Generation approach is very elegant and useful. It not only addresses the whole "business modeling" process, but it also provides a modern, simple and yet rather complete "framework" for describing your business on the general level. This has been missing until now. "Enterprise Architecture" is not really attractive to business people. They experience a disconnect between the rather elaborate specifications in such enterprise architectures and the reality that keeps business people awake at night (on bad days). This schism is really one of the very big problems of developing the business by way of deploying information technology. Traditionally such projects have been dealt with based on engineering approaches (with a reliability focus). These things are changing, now. Concept mapping (with a validity focus) is a perfect consort for the business models, the business plans, strategy trees and so forth.

In Chap. 5 we will look at where to find the concepts and the relationships and in Chap. 6 we will look at how to actually map the concepts and get going from there.

Let us review the business model generation approach combined with concept mapping:

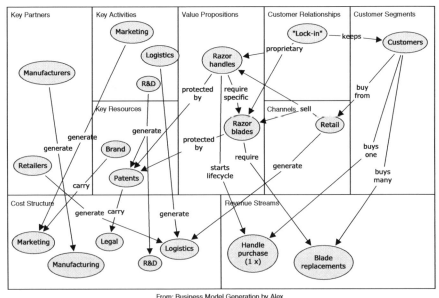

Fig. 5.1 The Gilette Business Model on top of the Business Model Canvas (Osterwalder and Pigneur 2010)

As you have seen before, the business model is elegantly expressed as a business concept map. All the concepts are pretty high-level, but that is fine. The business model is high-level by definition. The combination of the canvas and the concept map is a highly compressed representation of a quite complex narrative. If you tried to explain all of the meaning of the combined picture, it would probably be a 30-page document.

It is certainly very tempting to propose that you should draw a general level concept map as soon as you can, on the level of your business model. And then you probably should draw nine more – One for each building block of your business model. Keep them all as one-pagers.

5.2 How to Identify Business Concepts

As most people are aware, the key questions in most situations are: What? Where? When? Who? How? Why? And last, but not least: How much?

"How" can be broken down into two separate issues:
- How to do it?
- What are the rules for doing it?

The "How to do it" we will let the business process experts decide. The good thing about business processes is that you can always change and adjust them. The same can be said about the "Who" dimension – at least when one is talking about business internal organizational matters.

What the rules are is certainly interesting from a business information perspective. Business rules are sort of a bridge between the business information contents and the business processes. However, they are low level things, actually the lowest level of analysis. Most of the rules may be postponed to e.g. a system design process at some later point of time. Knowing when to stop at analysis time is therefore a very important skill, which you should think actively about.

The "Why" perspective is special in that it is essentially the link between the business model, business strategy, business plans and the business information needed to support those. To us (the analysts), they are sources of inspiration. And we might even comment on them and suggest improvements. But they belong to the business management, not to us.

Think in the context of "business objects", "actors", "events" and "measures". Together they look like this:

Perspective	Explanation	Examples
What?	The "business objects"; be they physical or abstract. May represent "things" you can kick and may also represent business events	Sales order, product, city, customer, invoice, financial statement, budget, all kinds of "Things"
Where?	Essentially another variety of business objects	City, branch, department, address
When?	Time and calendars in numerous disguises. Essentially also specialized business objects	Year, day, fiscal year, time of day and so forth
Who?	The "actors" of the business model expressed on the business object level, not the individual level	Customer, authority, business partner, vendor, employee and so forth
How much?	The "measures" which are necessary to conduct the business and to manage and control it	Invoice line, amounts, volumes, machine hours etc.

Note that "business objects" usually come equipped with descriptions. These may be texts and may also well be other describing properties, such as size, color, age and so forth.

All of the above are – in the concept mapping approach – concepts.

Obviously the concepts are interrelated. Customers place orders, Products have a price and so forth. In the concept mapping approach, these relationships are represented by the connecting lines between the concepts:

Fig. 5.2 A very basic concept map

Notice that this little concept map may be read as a sentence. The two concepts are subject and object, respectively. The relationship description is the "verb" (or the predicate, if you are into formal logic). This makes concept maps highly intuitive and understandable by most (who know what the business is about).

Let us move into where to find the knowledge about the business concepts and their relationships.

5.3 Concept Mapping Brainstorming Workshops

As described in Chap. 4, workshops are key to the success of the design thinking approach to business analysis using concept mapping.

The next parts of this chapter will point you to other sources of knowledge about the business concepts within your organization (and their relationships), such as Excel spreadsheets.

There are very good reasons why those ("passive") sources are not sufficient:
- The sources may not be representative (who knows?).
- The sources will not be complete.
- The sources could be out of date.
- The IT systems and business processes are being used differently than originally documented.
- The IT systems and business processes are being used differently in different parts of the organization.
- You cannot find the perfect future in elderly sources.

There may be reasons for going to sources other than people's minds, which is what we will look at the rest of this chapter. However, those kinds of analysis activities must be second priority because of the uncertainties laid out above.

First and foremost, your business driver for your project is very likely a requirement to change/improve/rework/create something. Using design thinking approaches. Workshops have proven to be furnaces of creativity and wisdom. (You can find several psychology books and papers discussing this). And they work particularly well, when you use CmapTools or similar products for brainstorming sessions.

5.3 Concept Mapping Brainstorming Workshops

Which particular style of workshops, you will need, depends on culture, your own style and the people present. See Chap. 4 for details about whom to summon and how the process may look like (going from Exploration over Ideation to Implementation).

There are several good books available on business analysis workshops. One of them is "Business Analysis Techniques – 72 Essential Tools for Success" (Cadle et al. 2010).

In addition to normal workshopping considerations there are some additional things to consider.

Following up is almost completely done using the concept maps, the concept definitions and simple ToDo-lists maintained from workshop to workshop.

We have had very good results with having both a "scribe" and a "facilitator" leading the workshops. The facilitator operates the concept mapping tools using a data projector, and the scribe catches (in text) all other notes and items for the ToDo-list.

Brainstorming works very well using concept maps. You may start from scratch or you may have prepared a few concept maps in advance (looking at the sources described in this chapter).

Concept maps replace Post-its and "talking walls" almost all of the time. There may be a few very difficult conceptual investigations where you might want to use some of the traditional techniques.

Concept maps replace mind maps and are even more efficient. (The real world does not look like snow flakes, but like a mesh of connected graphs.

Review Chap. 4 for a list of the kind of questions and observations you need to go through in each of the three stages.

The length of the workshops – and the number of them – depends on the size of scope of project, of course. Some small tasks may be handled in a couple of hours (from Exploration over Ideation to Implementation), whereas other (larger scopes) may take 1–3 workshops in each stage. Workshops that last longer than 3–4 h are not recommended. Neither the participants nor the facilitator can keep the tight focus required much longer than that.

In between the workshops, the analyst/scribe/facilitator have work to do:
- Documenting the results (e.g. making the concept maps look nice, editing and refining the concept definitions document and so forth)
- Preparing the next workshop (could be a couple of draft concept maps, drawn from passive sources like e.g. Excel and the like)
- Interviewing knowledgeable people in order to sort out uncertainties and open questions
- Looking for confirmation of e.g. concept relationships in actual data
- Reporting back to the project manager/owner and so forth.

As stated above, workshops with brainstorming (cf. Chap. 4 above) is the best source of conceptual knowledge of all. All other sources should be considered complimentary. They may be useful in the preparatory stages (before the workshops) and/or as sources of detailed conceptual information – once the core business concepts are defined and in place in business concept maps.

To emphasize the importance of brainstorming: During an ideation stage in a logistics company the context was stops on delivery routes. For various business reasons some stops were believed to be unreasonably expensive compared to the value of the payload delivered. It was decided to dig into the problem and a "location fitness index" was sketched (to be implemented in the ERP-system). This happened in a business intelligence ideation context and would probably not have happened by itself during the day-to-day operational activities.

Let us now look at where to find meaningful information about business concepts and relationships in passive sources. This is typically done as part of advance preparations and/or as part of detailed analysis and design in the implementation stages.

5.4 Excel: Where Meaning Lives!

Microsoft Excel (and some other spreadsheet products like OpenOffice) contain much of the knowledge about the concepts, you are using to run your business! It is as simple as that. Let us have a look – by way of examples – of how we can extract the concepts (and the relations) from Excel sheets.

We will start with a very basic spreadsheet:

5.4 Excel: Where Meaning Lives!

	A	B	C	D	E	F
1	Product	Customer	Qtr 1	Qtr 2	Qtr 3	Qtr 4
2	Alice Mutton	ANTON	$ -	$ 702,00	$ -	$ -
3	Alice Mutton	BERGS	$ 312,00	$ -	$ -	$ -
4	Alice Mutton	BOLID	$ -	$ -	$ -	$ 1.170,00
5	Alice Mutton	BOTTM	$ 1.170,00	$ -	$ -	$ -
6	Alice Mutton	ERNSH	$ 1.123,20	$ -	$ -	$ 2.607,15
7	Alice Mutton	GODOS	$ -	$ 280,80	$ -	$ -
8	Alice Mutton	HUNGC	$ 62,40	$ -	$ -	$ -
9	Alice Mutton	PICCO	$ -	$ 1.560,00	$ 936,00	$ -
10	Alice Mutton	RATTC	$ -	$ 592,80	$ -	$ -
11	Alice Mutton	REGGC	$ -	$ -	$ -	$ 741,00
12	Alice Mutton	SAVEA	$ -	$ -	$ 3.900,00	$ 789,75
13	Alice Mutton	SEVES	$ -	$ 877,50	$ -	$ -
14	Alice Mutton	WHITC	$ -	$ -	$ -	$ 780,00
15	Aniseed Syrup	ALFKI	$ -	$ -	$ -	$ 60,00
16	Aniseed Syrup	BOTTM	$ -	$ -	$ -	$ 200,00
17	Aniseed Syrup	ERNSH	$ -	$ -	$ -	$ 180,00
18	Aniseed Syrup	LINOD	$ 544,00	$ -	$ -	$ -
19	Aniseed Syrup	QUICK	$ -	$ 600,00	$ -	$ -
20	Aniseed Syrup	VAFFE	$ -	$ -	$ 140,00	$ -
21	Boston Crab Meat	ANTON	$ -	$ 165,60	$ -	$ -
22	Boston Crab Meat	BERGS	$ -	$ 920,00	$ -	$ -
23	Boston Crab Meat	BONAP	$ -	$ 248,40	$ 524,40	$ -
24	Boston Crab Meat	BOTTM	$ 551,25	$ -	$ -	$ -
25	Boston Crab Meat	BSBEV	$ 147,00	$ -	$ -	$ -
26	Boston Crab Meat	FRANS	$ -	$ -	$ -	$ 18,40
27	Boston Crab Meat	HILAA	$ -	$ 92,00	$ 1.104,00	$ -
28	Boston Crab Meat	LAZYK	$ 147,00	$ -	$ -	$ -
29	Boston Crab Meat	LEHMS	$ -	$ 515,20	$ -	$ -
30	Boston Crab Meat	MAGAA	$ -	$ -	$ -	$ 55,20
31	Boston Crab Meat	OTTIK	$ -	$ -	$ 368,00	$ -
32	Boston Crab Meat	PERIC	$ 308,70	$ -	$ -	$ -

Fig. 5.3 A data table in a Microsoft Excel spreadsheet

(The spreadsheet is downloaded from Microsoft's template website for Microsoft Office).

It is easy to identify the following concepts: Product, Customer, Quarter and Sales Amount (the numbers). Obviously there is some sort of relation between Product and Sales Amount, between Customer and Sales Amount and between Quarter and Sales Amount.

Let us move up the ladder a little. Here is a somewhat more complex spreadsheet:

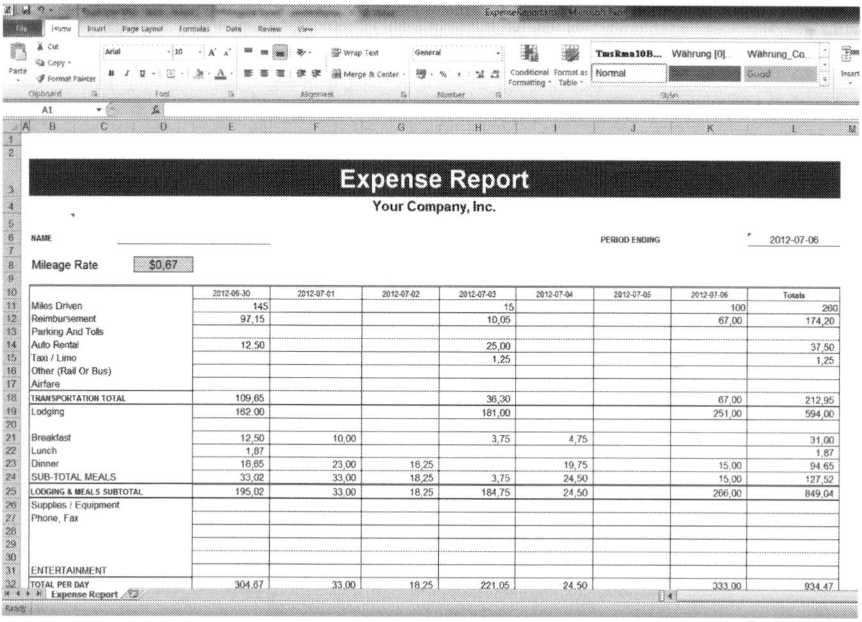

Expense Report

Your Company, Inc.

NAME

PERIOD ENDING 2012-07-06

Mileage Rate $0,67

	2012-06-30	2012-07-01	2012-07-02	2012-07-03	2012-07-04	2012-07-05	2012-07-06	Totals
Miles Driven	145			15			100	260
Reimbursement	97,15			10,05			67,00	174,20
Parking And Tolls								
Auto Rental	12,50			25,00				37,50
Taxi / Limo				1,25				1,25
Other (Rail Or Bus)								
Airfare								
TRANSPORTATION TOTAL	109,65			36,30			67,00	212,95
Lodging	162,00			181,00			251,00	594,00
Breakfast	12,50	10,00		3,75	4,75			31,00
Lunch	1,87							1,87
Dinner	18,65	23,00	18,25		19,75		15,00	94,65
SUB-TOTAL MEALS	33,02	33,00	18,25	3,75	24,50		15,00	127,52
LODGING & MEALS SUBTOTAL	195,02	33,00	18,25	184,75	24,50		266,00	849,04
Supplies / Equipment								
Phone, Fax								
ENTERTAINMENT								
TOTAL PER DAY	304,67	33,00	18,25	221,05	24,50		333,00	934,47

Fig. 5.4 A classic expense report with formulas in a Microsoft Excel spreadsheet

(The spreadsheet is downloaded from Microsoft's template website for Microsoft Office).

Looking at the spreadsheet (ignoring formulas for the moment) we are able to extract the following concept map from it:

5.4 Excel: Where Meaning Lives!

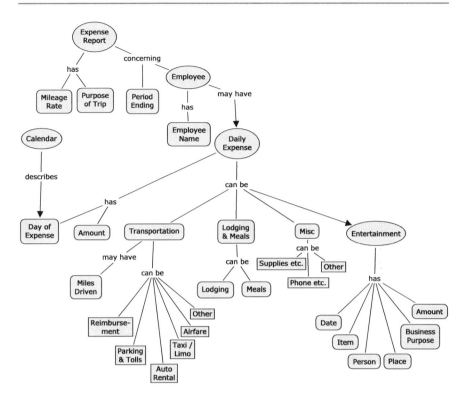

Fig. 5.5 The expense report as a concept map

What this tells us is not only a lot about expense reports, but it is also, on a more general level a description on our policy for travel expenses. The person, who created it (me), has been very mechanical in the sense that he has attributed "Mileage Rate" to the expense report. That is what it seems to be, but in reality, there is most likely a Mileage Rate for a period of time (from date, to date) until it changes. Also, the reimbursement amount for mileage is not clear. Checking the formulas, we discover that cells E12 thru K12 contain the formula: =IF(E11; ROUND(+'Expense Report'!D8*E11;2);""). Since D8 is the actual mileage rate (which the text in B8 tries to tell us), and since E11 thru K11 is the "Miles Driven" (text in B11), we can conclude that the cell references carry meaning. Namely that Reimbursement (text in B12) is a concept, which is calculated from "Miles Driven" times "Mileage Rate".

This knowledge enables us to redraw the concept map, where we correct the reimbursement stuff and also take into account that we investigated how the organization handles Mileage rates:

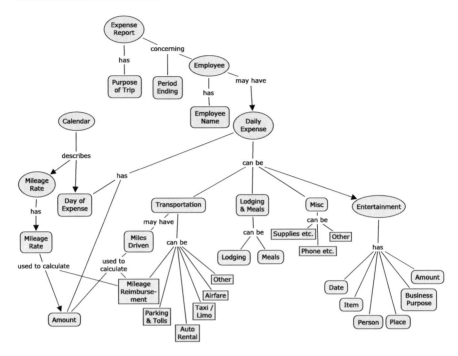

Fig. 5.6 The expense report extended with mileage rate concepts

Notice that Mileage Rate now is a business object in its own right – reflecting that the company has a list of Mileage Rates over time.

You don't have to be an Excel shark to figure this out, of course. The knowledge about how to calculate mileage reimbursement is everyday knowledge for most employees. You probably know a lot of cases, where much more complicated formulas in your Excel sheets actually provide relations between concepts. That is what is happening; visually we see some concepts materialized as texts (typically) in certain cells. And, behind the scenes, formulas establish a lot of relations between concepts. Some of the Excel functions – besides cell references, which may carry meaning like this, are:

- Sum (may sound trivial, but that is not always the case)
- Most of the database functions
- Many of the financial functions build upon a certain concept model, which you have to follow
- Many of the lookup and reference functions (such as VLOOKUP and so forth) implement relations between concepts
- Quite a few of the text functions (such as FIND, SEARCH, MID, LEFT and so forth) are used because there are implicit relations in the data in the cells.

In conclusion: Your most important and most often used Excel sheets are key resources for harvesting your key business concepts! Another way to say it is that maybe the reason for the Excel sheets to be plenty is that they (at least some of them) are meaningful!

5.5 The Chart of Accounts is full of Meaning

Most people think that Charts of Accounts are very mundane. In fact, they are full of meaning:

7100 Cost
7105 Cost of retail
7205 Cost of raw materials
8000 Operating expenses
8100 Building maintenance expenses
8200 Administrative expenses
8300 Computer expenses
8400 Selling expenses
8500 Vehicle expenses
8600 Other operating expenses
8610 Cash discrepancies
8620 Bad debt expenses
8630 Legal and accounting services
8640 Miscellaneous
8700 Personnel expenses
8710 Wages
8720 Salaries
8730 Retirement plan contributions

First of all there is the account hierarchy. In the (quite arbitrary) snippet of a chart of accounts above, "Cash Discrepancies" is in the group "Other Operating Expenses", which is in "Operating Expenses". If we opened up "Selling Expenses" (or any other of the Expense Account groups), we could find other interesting hierarchical classifications. (That is what we are talking about here). Sum accounts are concepts in their own right. Here is the concept map of Cash Discrepancies:

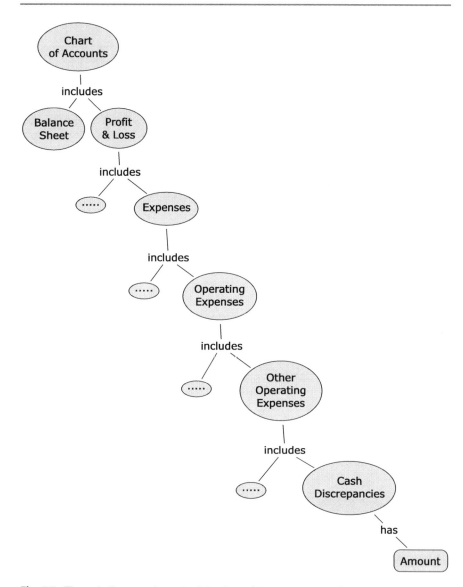

Fig. 5.7 The cash discrepancies path of the chart of accounts expressed as a concept map

Add to that the in real-life Charts of Accounts you find lots of both industry specific and company specific concept names. A Chart of Accounts for shipping companies, for example, contain a lot of shipping "lingo" in the account names: Charterers, Agents, Owners (ship-owners), Bunkers (fuel), Brokers, Bunker Swap, T/C (time charter, used on many P&L accounts), IFO, MGO and MDO (bunker types), Dead freight, L.O.W. (Last Open Water), Detention, Demurrage, Despatch, Ballast bonus, Dry-docking, Stevedoring, Ocean Routing and many more.

Certainly the finance department staffs have a point, when they say that they hold the key to understanding a lot of what the company is all about!

Let us turn that around a bit:

> Your business concept maps are to your business information asset what your chart of accounts is to your financial assets!

Accounting principles blur the picture, at times. How about the account groups "Overhead Costs" and "General and Administrative Costs". Both of them may well contain accounts for e.g. Rent, Utilities, Telephone, Travel, Consultants and what have you. Clearly the concept of "Overhead Costs" needs to be explained thoroughly and clear rules for posting of such costs should be specified.

There is also the issue of perspective. Accounting is concerned with the financial aspects of running a company. Obviously operating a company has many more perspectives depending on what the company does. These aspects are also important and you will have to look outside of accounting to find those concepts.

5.6 Applications and Databases Might Be Meaningful, Too...

Most people will expect that databases and applications might be full of useful concepts. In reality it varies quite a lot.

Let us begin with applications. There could be useful information about concepts and relationships in the texts on screens and reports, and there could also be good information in the documentation (user guides and such). However, consider these factors:

- The application is a standard system – some of the concepts are not quite as we understand them in our company
- The documentation is out of date
- We do not use the application in the way it was originally intended to be used by its designers – one example could be that we only use certain parts of it, and in "novel" ways

So you will have to be careful here. Involve people in the organization – in particular those people, who actually know what is being done. And these people are often key resources, which many people ask for help. You will need their managers' attention in order to prioritize their time.

What about databases then? Well, they often lack important information. The diagram below is an everyday (for IT people) way to describe a database structure:

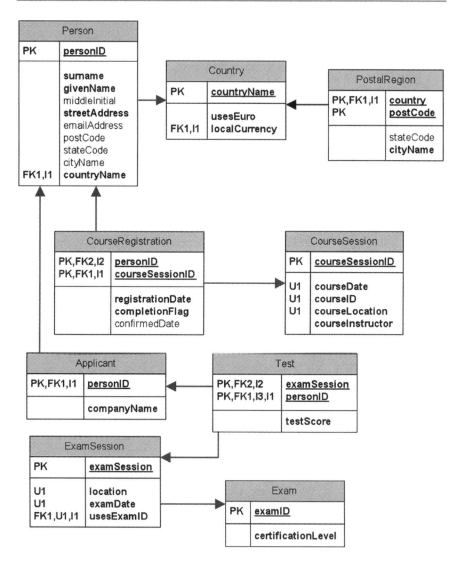

Fig. 5.8 A "logical" data model (entity-relationship style)

Being an IT construct it does look somewhat technical. Please ignore the "PK", "FK", "I1" etc. Seen from above (in the helicopter, if you will), we can see:
1. Properties of business objects (e.g. Person) are "bundled" together as a list of properties. In a concept map we would like to have the names of the relations between the concepts (e.g. that Person *lives* at StreetAddress). This meaning is missing. In other words, we cannot be sure whether the Address is an address of residence or of delivery, for example.
2. There are relations (indicated by the arrows, which – by the way – point the other way than what one should intuitively expect), but they are only between the business objects.

The detailed relations between concepts need to be found elsewhere. Add to that that many databases were designed either:
- A long time ago where it was best practice to use technical names instead of business terminology in databases (eg. "OHNUM" means Order Header Number, whatever that is), or
- By IT developers who really did not care about meaning and terminology (resulting in even worse naming than the example above).

Again, most often the documentation is not up to date. That means that you will have to involve IT, if you want to extract concept maps from database designs. As a starting point, you should ask for the Data Model. (Which ideally should be diagrams much like the one above supplemented with descriptions of the tables – containing business objects – and the attributes in the tables – properties of the business objects). Chances are that you will not get what you need.

Not everything about databases can be blamed on IT, though. Quite frequently you observe outbursts of creativity in the way business people use IT applications. Text fields are a good example. Do you, in your company, have text fields, which besides the text includes for example "X" or "23" to the extreme right of the text field? That is quite common. The meaning of the X or of 23 is of course, defined by the user – or a group of users – and could be "Check availability before invoicing" or "This is really something that should concern department 23". This is some of the issues, which we discuss more in Chap. 8 (about information quality).

One should not spend too much time looking at databases, when what you want to do is understand the business concepts. There are two key issues in this. One is: If you are using a standard ERP system, have you changed your business concept model to use the same concept model as the ERP-vendor has defined? And the other is: If you try to "re-engineer" concepts from a database design, you will have to look at the actual data. You must determine, if what you see in the database is what you would expect. This is something we in data warehousing do a lot, and it resembles archeology a lot...

5.7 Reports

Coming back to applications – the reports are most often of value. At least those reports which are being used week after week or month after month. And that brings us back to the spreadsheets, which really play a key role in understanding your business. A lot of reporting is done by way of spreadsheets.

5.8 Documents and the Internet Are Full of Meaning

Certainly. Your own documents and your own websites are loaded with the business concepts, which we are looking for here. But why stop there? What about emails? The Internet in general? Facebook? Twitter? Lots of good information out there.

We should be careful here. Scope creep is very easy to get into.

5.9 Take Control of What Your Business Means!

As demonstrated, the business concepts control the business processes. They are – in consequence – essential for a number of very important (and costly) activities and documents:
- Administrative guidelines, charts of accounts and many other documents used by staff frequently
- Spreadsheets – hundreds or maybe thousands of them – all of them have concepts spread out as row or column headers and in formulas, which really relate concepts together, when you think about it
- Engineering documentation of products
- Collateral of many kinds, including your websites
- Requirements and other specifications of IT systems
- IT data models in applications and databases
- Business intelligence reports and their "multidimensional cubes"

In brief, concepts control most of what is going on:

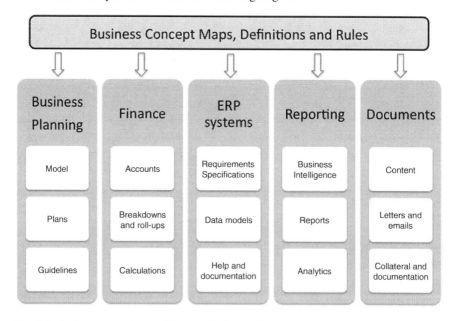

Fig. 5.9 Business concept maps "sit on top" of everything

Clear language and consistency is of essence. Just a few examples: What happens, if your instruction materials to the staff are using different concepts and rules? Or your accounting rules and formulas do not align with the business model? Or the data models in your IT systems are not aligned with the business (happens a lot...). Or your reporting is based on yesterdays business model (Seen that, anyone?). Or your websites are a mixture of defined concepts and something else? Confusion, mistakes and errors is what is happening!

In the long run the business environment will change and in order to keep up with that, the business must know what it is doing (i.e. understand the meaning of the narratives it uses to describe the world it is acting in and within). Some of the most obvious changes over the last 30 years have been those made possible by new information technology. This includes databases, ever more powerful hardware, nice graphics, interactivity a la the iPhone/iPad and so forth. That is not going to stop any day soon. As you will see in Chap. 10 (about business intelligence") the next wave – semantics – is making inroads into more and more companies. *Business concepts = business semantics!* The business advantages of semantics are many – cf. Chaps. 12, 13 and 14. Basically they boil down to being able to offer more intelligent value propositions in a turbulent sea of too much information.

Get busy working on mapping your business concepts today – you need them at hand! In section III of the book, we will look at a handful of brand new business opportunities, which become possible for you, if you "farm" and nurse your business concepts.

5.10 Business Dialects

What is of essence is that you understand the business. That is what the business concept map is about. The "common round" (the common understanding of meaning) in your company has some important differences in perspectives. They should be fully respected, because they are – most often – based on solid business reasons. In reality you will have the "Company Standard Business Concept Map", hopefully. But you will – for good business reasons – also have e.g. the "Bulk Market Company Standard Business Concept Map", and you also know that there are local "dialects" in Finance, Manufacturing and Marketing. The picture looks somewhat like this:

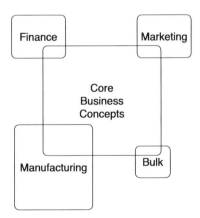

Fig. 5.10 Overlapping domains of business concepts and terminology

What we have in scope in this book are the core business concepts. And for harvesting them, you do not need to look at all documents, intranet pages, emails and so forth. The key elements will be strategic and tactic documents (business plans and the level below). Most of what we are looking for, is – actually – found more easily in Excel spreadsheets, reports and the other things, we have discussed so far.

This is true at the time of writing. However, there are very strong, new technological tsunamis coming upon us. Text analytics is a good word to look up, more about that later on in this book.

Now, given that you are in control of, and have full understanding of your business concept maps, what does that actually mean in terms of operating against competitors and on different markets? Obviously, there is potential here. If you know what you mean, then you should find out what your competitors mean (that may be accomplished by those forthcoming text analytic technologies). And you may also relate, objectively, to what the market(s) mean. This is the connection to Internet search, Facebook and Twitter.

This reinforces the fact that it is of essence to the business that it understands, defines and maps its business concepts. Doing so will facilitate business innovation: As part of the concept mapping process, using design thinking approaches, innovate conceptual solutions are created. At the same time it is a great way to reduce the time to market from the business specification (the concept maps) to the IT solution (using e.g. graph-level modeling such as the RDF-standard for semantics – cf. below in the third section of the book).

We have looked at where to find the relevant business concepts, so now we need to map them.

How to Do Concept Mapping 6

This chapter is educational in the sense that it instructs the readers in the very simple and highly successful concept mapping technique, which has been used by us since 2005 on a number of different projects for clients in both the private and the public sector. The technique is very pragmatic and interactive.

A **Business Concept Map** describes the Business Concepts and their Relations, which are used by the business on a daily basis, expressed in its own language in an intuitive way meaning that the business people can participate in the work of maintaining it.

The key idea is that the documentation of a business concept model shall be in a language that is both precise and intuitive and at the same time very close to the business. In fact the business people participate in the work and they are perfectly capable of maintaining the models themselves – Untouched by IT! We use a free concept mapping tool, CmapTools (cf. http://cmap.ihmc.us/ for more information). You can find information about how to install and operate CmapTools in Appendix 2. What we are describing here is the same kind of concept map diagrams, which were introduced in Chap. 2. The present chapter will be a step-by-step introduction to concept modeling, detailing all the (few) necessary steps in the process.

6.1 Concept Mapping Explained

The process of modeling business concepts is straightforward. Initially a facilitator (could be a business analyst) interviews a few representative business users to get an overview. At the same time he/she also asks for samples of reports, memos, spreadsheets, documentation of IT systems, screens and so forth – of relevance to the subject area in focus. The analyst then produces one or more draft concept maps (using e.g. CmapTools) based on the "harvested" concepts in all those documents.

Selected business representatives are then gathered for one or more 2–3 h workshops (brainstorming sessions, really) with the purpose of actually producing an initial version of the business concept map. It may look like this (intentionally very pedagogical example):

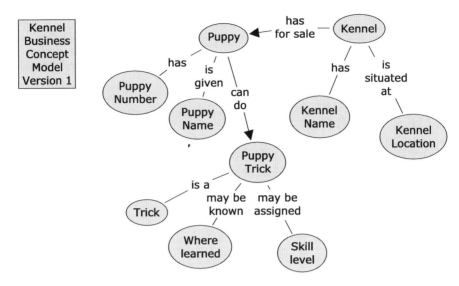

Fig. 6.1 Early version of a kennel business concept map

This is a very simple representation including:
- Business concepts,
- The relationships between them as well as
- Indication (by the use of arrowheads) of one-to-many relationships

Nothing more. It is important that the terminology is really the day to day language of the business. Also remember that the diagram can be read in sentences (e.g. Puppy can do Puppy Trick), which is very nice for verification purposes when you sit in a brainstorming session. All those little "sentences" are called propositions in CmapTools, and they are one of the central ideas of concept mapping.

Mapping rule:

▶ Use circles for all concepts in the beginning – when you are brainstorming or sketching the thing.

Mapping rule:

▶ Name all relations between concepts.

Mapping rule:

▶ In the beginning – be relaxed about using arrows (one-to-many relationships) or not; you can always come back and add that detail later.

6.1 Concept Mapping Explained

The business concept model is further refined in subsequent iterations on workshops and using other potential feedback mechanisms such as emails, intranets etc. The final result could look something like this:

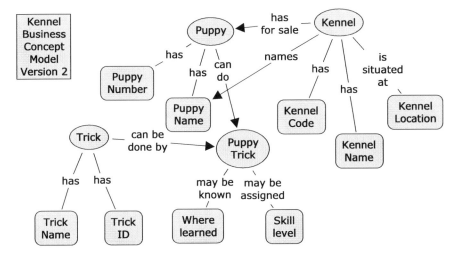

Fig. 6.2 The kennel business concept map extended with properties of the business objects

More concepts have been introduced and the "Tricks" have been separated. All the concepts in circles denote business objects (see definitions in the following text).

Notice that in this second version of the concept map a new type of icons (rectangles with rounded corners) is introduced. They are used to denote properties. In other words business concepts represent either business objects or properties of business objects.

Mapping rule:

▶ Use circles to denote business objects.

Mapping rule:

▶ Use rounded rectangles to denote properties (of business objects).

There is a slight risk of missing some things in this approach. Note that "Kennel Location" (in the model above) is probably not just a property, but is more likely to be a complex geographical concept model, such as a postal address with city, post code, street etc. However, you have got to stop somewhere! So, the geography was probably deemed not interesting (out of scope) and simplified to a simple text property for the time being.

6.2 What Are Business Objects?

They are basically all the "things" that the business works with, including "agents" (people or organizations) and "documents" such as invoices and so forth.

The car rental example, which we looked at in Chap. 2, had the following basic business objects: Branch, Customer, Booking, Car and Rental. Some additional business objects are: Car Group, Car Model, Club Membership, Discounts and Upgrades and even "Bad experiences"!

6.3 Properties of Business Objects

Properties are the concepts, which we use to describe our business objects. One example is the concept of Color, which could obviously be a property of Car. However, the EU-Rent car rental company is not at all interested in colors. In their perspective the following is what is necessary to know about cars:

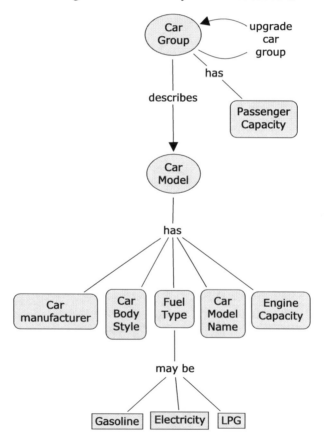

Fig. 6.3 Car concept map from the EU-Rent context

Notice all the properties, which describe the rental cars. Also notice that in the diagram above "Fuel Type" is specified as having three (possibly more) possible values: Gasoline, Electricity and LPG (Liquified Petroleum Gas). The icon used for that is squared rectangles. This may be a good idea to use for explaining what you are talking about in a Concept Model. But do not use it all the time. In the example above it makes sense to be specific, because not everybody would think of electricity as a fuel type. That is what EU-Rent has decided it is – in their case.

Mapping rule:

▷ Use squared rectangles to denote actual values (of data) for pedagogical purposes.

6.4 Definitions and Other Specifications

What remains to be done is a simple document listing all the concepts and providing textual descriptions such as:
- Definition
- Description and comments
- Type
- Special rules
- Sample values

This applies to both the concepts themselves and to the non-trivial relationships between them. Here follows an example based on the Car Details concept map, which was presented just above. The table below is a simple "Business Vocabulary" offering definitions and examples of the concepts in use (in the EU-Rent example). Use Microsoft Word or Excel or similar for these textual descriptions.

Concept name	Definition	Type	Comments and examples
Car Group	Different models of cars are offered for rental, organized into groups which establish a price point	Text	Family
Upgrade Car Group	The Car Group has an Upgrade Car Group that is used when no car of a requested Car Group is available	Text	
Passenger capacity	The count of adults, including the driver, that the car can comfortably hold	Number	
Car model	Cars of a given model are all built to the same specification, e.g., body style, engine size, fuel type	Text	Ford Fiesta
Car manufacturer	Producer of cars that EU-Rent has decided to do business with	Text	Ford
Car body style	Classifies a Car Model based on industry defined criteria	Text	Sedan, coupe, convertible

(continued)

Concept name	Definition	Type	Comments and examples
Fuel type		Text	Gasoline, electricity, LPG
Car model name	EU-Rent bases its model names on those assigned by the car manufacturers, but sometimes has to extend them to distinguish models with different engine sizes and numbers of doors	Text	Ford Fiesta 1.6 4-door
Engine capacity	Indicates the engine cylinder capacity in cubic centimeters	Number	

The power of examples is again clear, when you look at Car Body Style. People will think – "Aha, that is what they are talking about", when they see Sedan, Coupe and so forth. As you can also see, some of the definitions are not obvious. For instance: "Passenger Capacity" is a count of adults. Who would have thought about that? Businesses and organizations are full of little quirks like this. And if we can bring them out in the open, that is good.

As already mentioned, the authoritative book about how to work with definitions is "Definitions in Information Management" (Chisholm 2010). It is an excellent book, which is highly recommended as a companion to this book.

The business vocabulary documents – and the concept maps – are core business documents, which should be published internally on your intranet, for example. This ensures that everybody in the organization use precise and well-defined terminology – one of the key benefits of business concept mapping.

6.5 Structured Concepts

As you can see in the diagrams we use connecting lines to describe relations between concepts. Relations are important. They add structure to the concept map. Generally there are three types of relations:
- One-to-one
- One-to-many
- Many-to-many

One-to-one relations are found in two situations:

1. Between *a business object and one of its properties* (normally a business object has only one occurrence of a property like e.g. Engine Size, but there are exceptions).

6.5 Structured Concepts

2. Between *two business objects*, like e.g. a EU-Rent Operating Company (e.g. EU-Rent Sweden) and its' Insurer (the insurance company that serves the EU-Rent Operating Company).

Here is another example of a property of a business object:

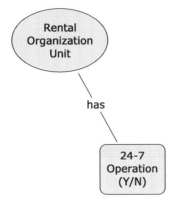

Fig. 6.4 The 24-by-7 property of the rental organization unit

In other words "Rental Organization Unit" (probably a Branch) either has 24 by 7 operations or not (still considered a one-to-one, mind you).

And here is the Insurer example:

Fig. 6.5 A 1:1 relation between two business objects

One-to-one relations between business objects are not so frequent, but they do happen.

Mapping rule:

▶ Use a connecting line without an arrowhead to denote a one-to-one relation between concepts.

One-to-many relations are the normal situation between business objects. The example concept map first presented in Chap. 2 is full of them:

6 How to Do Concept Mapping

Fig. 6.6 Overview of the EU-Rent car rental business

Remember that we use arrowheads to denote the "many" side of the relation. Strictly speaking, one-to-many really means: *A business object may have zero, one or more relations to another business object.*

This means that a Customer (in the example above) may have no Rentals and still be a Customer (at least that is what the diagram says). If this is not the case, it should be documented in the definition of the Customer concept (in other words: what are the rules for when we can call somebody a Customer – these may be quite complex in the real world).

Mapping rule:

▶ Use a connecting line with an arrowhead at the "many" side to denote a one-to-many relation between two business objects.

Now, what about many-to-many relations? On the surface it seems as if it should be a pretty common thing to happen. It is indeed true that there is a many-to-many relation between Customer and Branch, for example. Any one Customer may have visited many Branches and certainly (hopefully) many Customers visit each Branch. However, in the real life, what happens is that we (the business people)

6.5 Structured Concepts

introduce something in order to record that relationship. In the car rental business that "some thing" is the Rental (agreement), cf. the diagram above. In other words, in almost all cases, the many-to-many relation materializes as something concrete. Probably a human construct like an Agreement or similar, but nonetheless a real business object in its own right.

Should you run into a situation, where you cannot name "the thing in the middle", here is how you may choose to diagram it:

Fig. 6.7 A many-to-many relationship between two business objects

Mapping rule:

▶ Carefully check whether there really is no business object in the middle of a Many-to-Many relation. If it is really missing, use a connecting line with an arrowhead in each end (tip: Use the "Connection direction" part of the Style Palette of CmapTools).

The names of the relations are important. Quite often, you may use one of the general relationship types such as:
- "Has", implying that something "owns" something else, or
- "Describes", implying that something is described by something else (usually Properties)
- "Consists of" or "is divided into" or "includes", implying that something is a collection, upper level or generalization of something else
- "May be", implying that something may be one out of a list of some things else.

But in some cases (and there are more of them than one should think) a more specific word (typically a verb) is relevant.

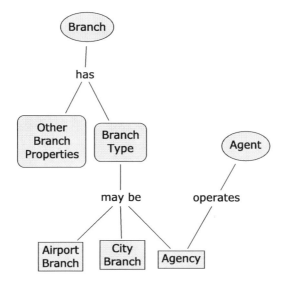

Fig. 6.8 A named relationship with a business-specific meaning

In the diagram above, the Agent is "operating" Branches of the Type "Agency".

Such specific relation names are quite important and you should make an effort out of trying to give them descriptive names.

Finally, in some of the diagrams above, you have noticed that we "bundle" relations of the same kind together. So, instead of having this:

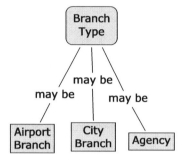

Fig. 6.9 Many relationships with the same meaning

Then we do it like this:

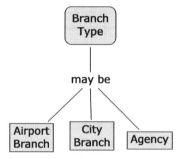

Fig. 6.10 Many relationships with the same meaning reduced to one

Modeling rule:

▶ Bundling together relations of the same type makes diagrams easier to read.

6.6 Layout of Concept Maps

There are just a few guidelines for producing "nice" concept maps. Many people instinctively look at things top-down and left-to-right (but there are cultural variations on this). Consequently, complying with this psychological fact makes the diagrams easier to read.

Modeling rule:

▷ Arrange concept maps so that they should be read top-down, left-to-right (or according to your own cultural conventions). This will facilitate a more intuitive understanding of the Concepts and the structure.

Everything should not be crammed into one page. Instead you should organize your diagrams with not more than 20 concepts (max!) at each page. A convenient way to help you to organize is to use a visualization like this:

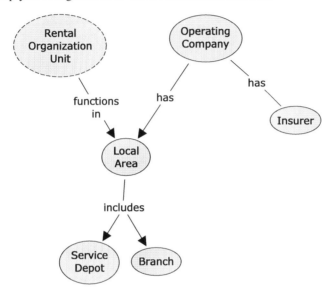

Fig. 6.11 Using dashed contours to visualize "off-page concepts"

The "Rental Organization Unit" in the diagram above is dashed, which – by convention – indicates that you should go to another diagram to find more details about it.

6.7 Dealing with Logic

What is differentiating business rules from concept maps? There are two dimensions in the answer:
1. Data
2. Logic

Data is commonly referred to in business rules (as e.g. "It is obligatory that the duration of each rental is at most 90 days"). Normally we do not go down that deep in the concept maps diagrams. There are times, however, where we do it – to be pedagogical as in this little (fragment of a) concept map:

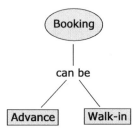

Fig. 6.12 Visualizing (instances of) "data" in concept maps

Notice that we use a different visual object for "data" than for concepts.

Sometimes it helps the reader understand the concept, if there are a few examples of the actual values, which are relevant.

The issue of logic is a very big issue. When we do concept mapping – as described in Chap. 6 – we do not care too much about logic in the diagram. Concept maps are not about logic; they describe concepts and their relationships. In business rules, on the other hand, you typically have a lot of logic. The little diagram above can be expressed as a rule saying: "A booking (of a rental of a car) can be either an advance booking or a walk-in booking".

Business is indeed performed by people and cognitive scientists today agree that logic and rationality is only a piece of the picture of human cognition. A good example is the (systematic) errors most people make when reasoning with probabilities. Add intuition, gut feelings and so forth; this complex set of capabilities is what we are and this is what we should support.

6.8 When to Stop?

You should be aware that there is a limit to how far you can take business concept modeling. Let us be concrete here. In the EU-Rent case study we find this complex business rule:

> Each rental booking has exactly one booking date/time If the Rental is a Points Rental (Membership Points) the booking date/time must be <u>at least 5 days before the scheduled pickup date/time of the rental</u>.

In other words, if you rent a car using membership points, you should book at least 5 days in advance. This could be diagrammed like this:

6.8 When to Stop?

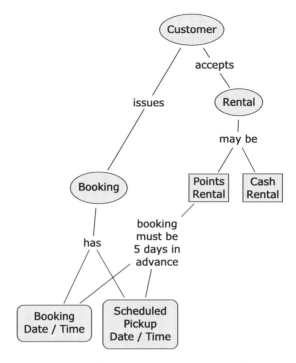

Fig. 6.13 A business rule about bookings of points rentals expressed as a concept map

This is of course in itself fine. But, how many of such detailed rules do you have? And imagine how your concept maps are going to look if you went into all that detail? So we have to stop somewhere. And the golden rule for stopping is that if you get into having actual data on the diagram, do not do it. In this case the actual data is "5 days in advance". That is a business rule and if you want to document it, please do so in the textual descriptions. Another alternative is to get into a project based on business rule software. They do exist and they do provide value, if you have many, complex and volatile business rules. We will get back to business rules software later in this book. Most often the recommendation is to document them in Word or Excel as they surface.

A more subtle explanation of the difference between business rules and semantics is: "If it further defines or extends the meaning, then it is semantics. If not, it is a business rule". In the example above it does not really matter whether a points rental must be booked 5 days or 7 days in advance. It remains a points rental, whichever rule is applied.

Modeling rule:

▶ If you find that you are including data in the diagrams, do not do it. Detailed rules should be in the textual descriptions, not in the diagrams.

Very important mapping rule:

▶ Concept maps must be approved by the business!

6.9 The Concept Mapping Tool Par Excellence

Is CmapTools. It is the tool we always use for drawing the concept maps. It is based on theoretical work done in the 1980 and the 1990 by Professor Joseph D. Novak (at that time at Cornell University). The tool, CmapTools as we know it today, came to life around year 2000 and it picked up speed in the educational sector a couple of years later. The publisher of the software is the Florida Institute of Human Machine Communication, IHMC, where Prof. Novak works.

Refer to Appendix 2 for a short "user guide" to CmapTools.

6.10 Real Life Examples of Concept Maps

On the following pages you will see examples of concept maps from the real world (generalized in order not to reveal client information). They have all been used in data warehouse/business intelligence contexts. The examples – as presented here – are slightly simplified and are intended to help you get the picture: Concept mapping is a way of understanding your own business that creates real value for you.

6.11 General Company Structure

The following concept map is a generalized "mixture" of company conceptual structures from different clients:

6.12 Shipping

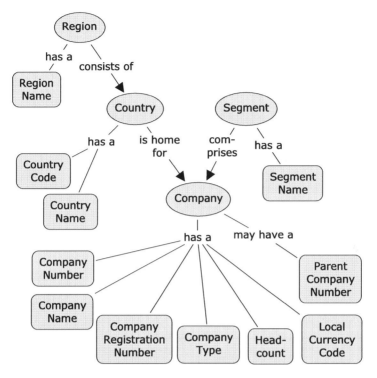

Fig. 6.14 A generalized (and simplified) company concept map

6.12 Shipping

In shipping there is something called the Contract of Affreightment (look it up in Google). It is basically the contract for lifting cargo. The diagram below is a snapshot from work in progress as part of a business intelligence effort.

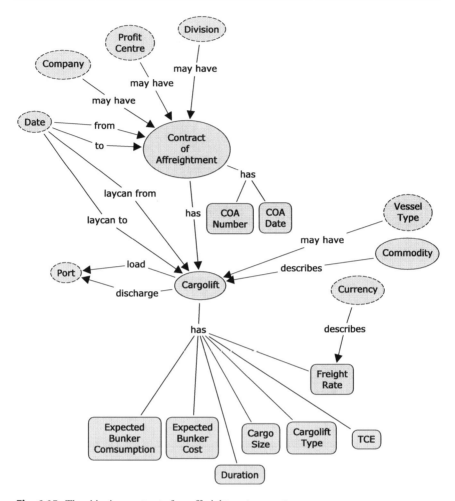

Fig. 6.15 The shipping contract of an affreightment concept map

6.13 Property Management

The diagram below is also a snapshot from work in progress (in a data warehousing setting). The concepts in the diagram relate to the complicated relations between Customers, Owners and Ownerships – comprising the Portfolios, which the property management company manages.

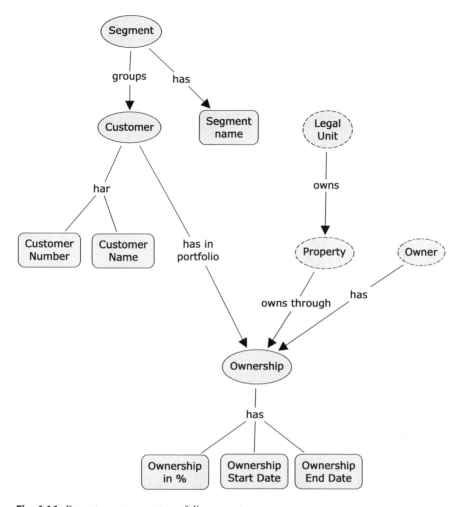

Fig. 6.16 Property management portfolio concept map

6.14 Car Dealership

The following diagram is from a car dealership (the context is a data warehouse project). It describes the information needed for particular vehicles.

6 How to Do Concept Mapping

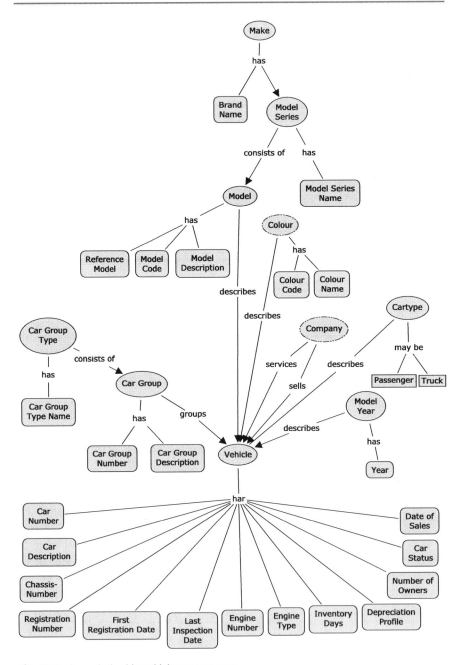

Fig. 6.17 A car dealership vehicle concept map

6.15 Public Sector Example

NASA has a site with over a 100 concept maps describing space exploration. See the whole work here: http://www.nasa.gov/exploration/whyweexplore/cmap.html

6.16 Concept Harvesting

Is your organization willing to be an early adapter of technology? If so, there are a number of emerging technologies and companies, who may help you to "harvest" concepts. That could potentially save you a lot of time in building your concept maps. You will not – any day soon – have a complete collection of business concepts "automagically", but "harvesting" them by machine can be a good starting point in selected areas.

The sources of business concepts are potentially the same sources as described in this book. So you let loose the "concept harvester" on:
- Business documents on shared directories and/or in content management systems
- Intranet pages and documents
- Databases
- Your home pages
- Sales, marketing and engineering literature
- Product documentation
- Competitors websites
- Social Media (FaceBook, Twitter ...)
- The Internet in general ...

There are different approaches to the basic technology. Look for things like:
- Text analytics (text mining)
- Content extraction
- Tagging/annotation
- Information/entity extraction
- Natural language processing
- Coreference
- Sentiment analysis
- Pattern recognition (in texts and elsewhere)
- Ontology building/construction/extraction

The field is moving rapidly, so here is some advice on what a "concept harvesting engine" should deliver:
- The basic "concept – relationship – concept" structures (e.g. "Customer issues Booking")
- The collection of the concept "triples" is essentially a graph resembling the concept map(s)

You will also need frequency counts and other statistics in order to determine whether a relationship is of high quality or maybe just a more loose relationship (just a little bit more that a coincidence).

There are several research-level projects in this area. A few of them aim at directly producing concept-maps as the end result. But they are not commercialized, yet (2012). You will need to be careful, if you have a case for a project of this kind.

6.17 Standard Business Concept Definitions

Another way to harvest concepts is to look at publicly available structured vocabularies etc. One of the most interesting is schema.org, which also has a number of concepts defined relevant for private enterprises. But there are many more standard concept schemes.

Let me give an example from schema.org: As part of the details of the Offer (offer to sell something) they have defined:

Concept name	Definition
price	The offer price of the product
priceCurrency	The currency (in 3-letter ISO 4217 format) of the offer price
priceValidUntil	The date after which the price is no longer available

Maybe this puts you into a dilemma. Should you adopt the concept names given by schema.org or not? The major search engines of the world know the naming scheme of schema.org. The sponsors of schema.org are, by the way: Google Inc., Yahoo Inc., and Microsoft Corporation.

It is not untrue that "Price Valid Until" is a pretty good name for a concept. And if you want some of your data (eg. your product catalog) to be searchable based on these standard names, you will have to either:
- Adopt the schema.org names and definitions to a certain extent, and/or
- Map your own concepts to those of schema.org

Much the same can be said about the plentitude of publicly available ontologies and vocabularies for very specific types of industries. At some point of time you will have to either change your vocabulary or to map you vocabulary to some industry standard. In either case your business concept maps come in handy, again.

This concludes the sections of the two powerful techniques of concept mapping and design thinking. The last section of the book is about taking advantage of having mapped and carefully designed business concepts readily available.

Part III

Business Innovation Using Mapped Business Concepts

"The world is but a canvas to our imaginations." Henry David Thoreau (1817–1862)

Concept Mapping and the Next Generation IT Paradigms 7

The term "map" in "concept map" should be taken almost literally. The concept maps map the concepts and relationships. On top of that they may be used to establish a map from the business level concepts and relationships to the IT systems, where the concepts and relationships are represented. (If they are supported by IT applications, that is):

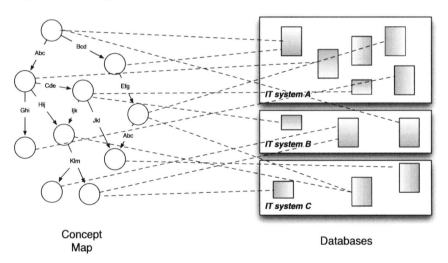

Fig. 7.1 Concept maps map business concepts to data models in databases

This detailed level of cross-referencing is, in fact, part of an enterprise level effort to manage the information assets of the enterprise. Today that discipline is typically called EIM, Enterprise Information Management, and it is right on the borderline between business and IT. It is a very big subject and it is excellently described in a number of books. See for example: "Making EIM Work for

Business" (Ladley 2010). Concept maps work very well in that context, too. John Ladley recommends simple "taxonomies" for EIM. We recommend simple concept maps. The two do very similar things. However, concept maps are not limited to hierarchical structures (as are taxonomies in principle) and concept maps communicate extremely well – in all kinds of organizations. This is because of the named relationships making the concept map a collection of sentences.

For very many years IT-people have come to expect that the normalized entity-relationship model or its sister, the UML class model, are best practice for data modeling. Information modeling outside of databases was left to library scientists with their thesauri, to information architects of websites with taxonomies and to knowledge managers with their ontologies.

Surprisingly, today a number of emerging paradigms in IT are *not* based on classes or entities. Here are eight of the more important ones:

- Information quality/master data management (MDM)
- Information valuation ("Infonomics")
- Business intelligence hierarchy management
- Business rules (automation of)
- Web 3.0 and semantic technologies
- Open information sharing (aka. linked data, incl. linked open data)
- Pull instead of push
- The proposals for the next generation of data management (NoSQL, Big Data and graphs)

We will look at each of those in the following, but overall the big differentiator is this fact: In all those eight paradigms it is necessary to work with named and typed relationships between all kinds of concepts, including the "atomic" properties of objects. As you have seen in the previous chapters, this is precisely the level of granularity of concept maps.

In essence there is an almost 1:1 mapping of concept caps to the structures used in all of those big opportunities for business innovation. You will have to provide more details in order to arrive at an IT solution, but the concept model stays untouched and is an integral part of the resulting application.

The opportunities sort of go together in a natural "roadmap" of where business concept mapping can take you:

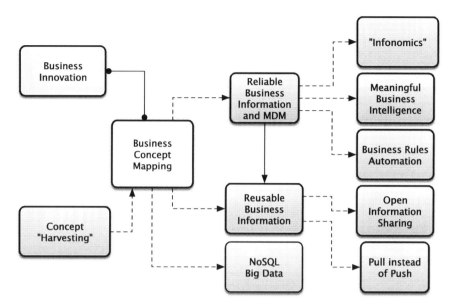

Fig. 7.2 Roadmap of the opportunities created by business concept mapping

Consider these opportunities as suggestions of how to re-architect a few of "the floors" of your enterprise "building". Having some of these suggestions in the back of your head as you go through the exploration-ideation-implementation cycles together with the business representatives helps you create new, real business value as you go along.

Have a look for yourself in the following chapters! What is important to realize is this: The preferred level of representation of business information is from now on the level of concepts and relationships, not the engineered approach from IT using tables with primary and foreign keys.

Opportunity: Reliable Business Information and MDM

Since around 2001–2002 there has been a growing emphasis on the issues of information and data quality. The five most important challenges, which gave rise to the awareness of the importance of this are:

Quality challenge	Description	Example
Consistency	Conflicts between terminology, definitions, representations, data values etc. across divisions, departments and IT systems	Different opinions on what "Cost Price" is
Completeness	Missing information	Product category only available on less than 50 % of the products sold
Accuracy	Incorrect information when compared to reality	Product weight registered as 10.5 when it should be 11.2
Validity	Information is in violation of expected and/or specified ranges or business rules	Salary is three times higher than the average for that type of employee in that department
Uniqueness	Duplicates exist. Can be either "same-same" or (worse) same identification but different content	Reuse of an old UPC for a new product (even though the old product is still found in the data warehouse)

A reliable business depends on reliable information

8.1 Data Profiling

Today a number of IT-supported solutions exist. The most noticeable is called "Data Profiling". Data profiling tools may be standalone solutions but most of them are built into a package of software products, not least database product packages. What you will get is a data profile that sits on the level of tables and columns in databases. Which is fine, if you are an IT developer, who needs to move data somewhere else and need to understand the data before you start programming (not a bad idea, to be recommended).

However, the business value of information quality needs to be on the business level. In the different IT systems there might be different problems with e.g. consistency, validity and even duplication across multiple IT systems. There might also be different representations (data models and so forth) in different systems, which further complicate the picture. What the business needs to know is e.g. a consolidated statement of information quality for customer information across different business units.

That is where your business concept maps and your definitions of the concepts come to the rescue. They are all and everything that the business has to put in front of information quality:

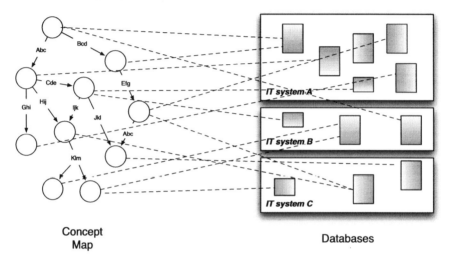

Fig. 8.1 Concept maps map business concepts to data models in databases – enabling you to manage information quality on the business level

Remember that the business concept maps are to your information asset what the chart of accounts is to the financial assets! Business information quality measures should be collected on the business concept level in order to give business value.

8.2 Master Data Management (MDM)

Master Data Management (MDM) is also a data quality initiative, which is strongly recommended. MDM is a combination of workflow (software supported) and "master data repositories" surrounded by a variety of master data services for delivery of managed master data to applications and databases (such as e.g. a data warehouse). One of the important facilities in MDM is "hierarchy management", the business of managing the roll-up and drill-down structures within the business information. Here is an example of a hierarchy, expressed as a concept map (naturally):

8.2 Master Data Management (MDM)

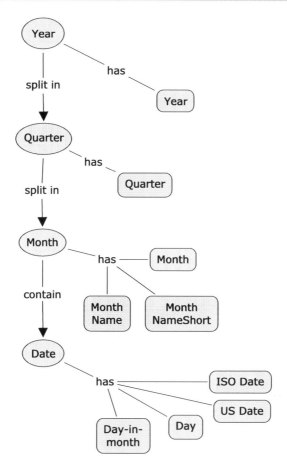

Fig. 8.2 A "calendar" hierarchy

In Microsoft SQL Server 2012's Master Data Services (MDS), for example, there is a facility for performing hierarchy management. The level of specification is called "terms", i.e. concepts. You actually need the concepts and their relationships in order to fully understand all the hierarchies. In many databases there are so-called functional dependencies (the data normalization jargon for hierarchies) hidden in (de-normalized) tables. Chances are you do not discover the relationships before you map the concepts and/or profile the data.

See for example David Loshin's Morgan Kaufmann OMG Press book (Loshin 2009) for more information about Master Data Management.

Information quality, master data management and business concepts maps go hand in hand. As the information quality expert Larry English so often underlines, definitional and structural quality are at least as important as data value quality (English 2009). In other words:

> It is not only the quality of the data itself, but also the quality of the concepts, relationships and definitions of the data, that matters.

Doing concept maps certainly improves the definitional quality dramatically. The authoritative book on information quality is: "Information Quality Applied", (English 2009). Highly recommendable – very detailed and very complete.

These days many of the data profiling tools offer support for "data quality scorecards" and dashboards with key performance indicators (KPIs). This is recommendable. However, KPIs on data quality should be on the *business* level, not the database level.

Opportunity: Information Valuation 9

"Infonomics" is a name for a proposed new discipline for valuing information as a corporate asset. The term was coined by Doug Laney, who is now at Gartner Group (Laney 2012). People, who have worked with the information asset in many companies, e.g. as an information management consultant, have seen the degree of impact the information resource has on business operations.

The balance sheet is full of intangible assets (copyrights and much more). Why not information? As Doug Laney (and John Ladley) rightly points out, one of the key characteristics of an asset (in accounting terms) is that it has probable future economic value even before it is used. See Gartner research note G00227057 (Laney 2012) for more information.

It will, no doubt, require yet some more hard work to get information on the balance sheet. However, as said before, the business concept maps are to the information asset what the chart of accounts is to the financial asset. Concept maps describe the business level, the data models in the databases are technical:

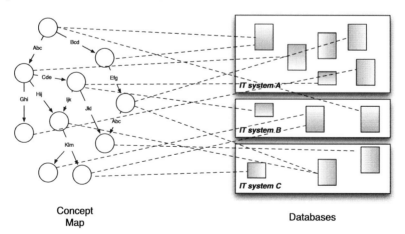

Fig. 9.1 Concept maps map business concepts to data models in databases – enabling you to manage information asset valuation on the business level

It does indeed make a lot of sense, that if you are going to perform a valuation of this asset, you must understand it. The business concept maps are there for that reason.

Opportunity: Meaningful Business Intelligence

10

One of the first books about business intelligence was "The Data Warehouse Toolkit" (Kimball 1996). Excellent book – much of the content is still valid. (There is a second edition from 2002). On the accompanying CD (!) there were some sample documents. One of those is a "Dimensional Model" (prepared by Ralph Kimball, Laura Reeves, Margy Ross and Warren Thornthwaite in 1998). One of the purposes of such a document is to describe the dimensions in the multidimensional model. For this purpose they employed simple diagrams as the one below:

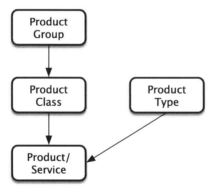

Fig. 10.1 A Kimball style product dimension diagram (Kimball 1996)

The example is obviously a Product/Service dimension containing two hierarchies: One on Product Group and on Product Type. This looks strikingly like a piece of a concept map! However, there are no named relationships between the concepts. Could be because that was not so easy to do in Visio at that time. This diagramming style is what got this author started on concept mapping. Initially using Visio (before and after it was acquired by Microsoft) as the drawing tool and switching to CmapTools, as soon as it appeared. A diagram like this is an implementation level business concept model – part of a specific design of an IT-solution (a data warehouse multidimensional data mart, to be precise).

Business intelligence shares the interest in hierarchies with Master Data Management (cf. above). And the hierarchies almost "fall out" of the concept maps – for "free". This is a considerable benefit of concept mapping. The straightforward re-use of the concept relationships is further underlined by the nature of the master data management or business intelligence hierarchy manager components in the various software configurations. In Microsoft SQL Server (since 2005) the Analysis Services component contains something originally called "Unified Dimensional Model (UDM)". As the illustration below depicts, one of the key things to specify in such a model is "attribute relationships". In fact, all the UDM concepts, which are shown with a thicker outline in the concept map below, are directly and readily available in the business concept maps:

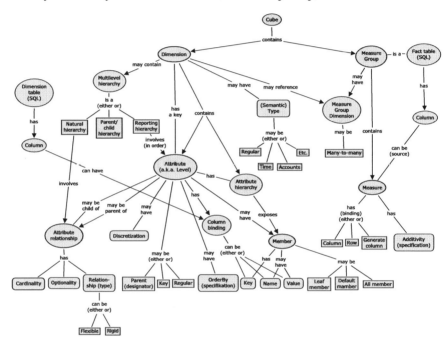

Fig. 10.2 A concept map of Microsoft SQL Server Unified Dimensional Model (UDM) (Mapped by the author)

The attribute relationships correspond – one-to-one – to the relationships (connecting lines) in the business concept map. And much of the rest is for use only in more special cases. Concept maps front-end business intelligence dimensions very well indeed. Today UDM is simply called multidimensional projects in the newer versions of Microsoft SQL Server.

As you have seen above, dimensions and hierarchies "fall naturally" out of concept maps:

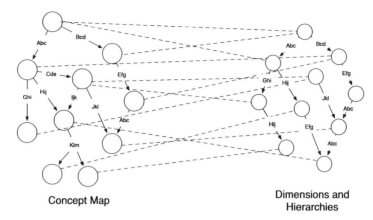

Fig. 10.3 Concept maps as front-ends to business intelligence dimensions and hierarchies

If you do not use a multidimensional approach but instead have direct access to databases, it is important to have a layer of managed business concepts in front of the databases:

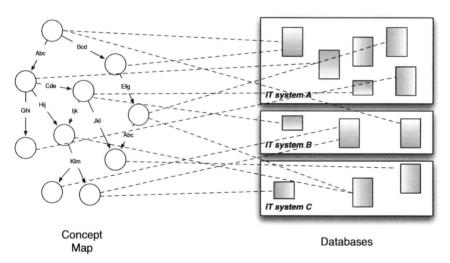

Fig. 10.4 Concept maps map business concepts to data models in databases - enabling meaningful business intelligence

Business intelligence must be based on the recognized business concepts to give meaning to the business. Many, if not most, databases do not use naming, which is immediately recognizable by the business users. A mapping from the database level to the business level must take place. The business concept maps define the target of that mapping.

Opportunity: Business Rules Automation 11

11.1 Concept Maps Versus Business Rules: Revisited

What happens when you add logic to a concept map is essentially a detailed specification of a rule that should be followed by people and/or systems. There are different forms and styles for different usages and/or different communities:
- When IT people do it, they typically call the result a data model and they use an engineering style diagramming convention called UML class diagrams (or an older style called Entity-Relationship diagrams).
- When library scientists (or archivists) do it, they might call it a taxonomy, a facet plan or an ontology.
- When knowledge managers do it, they also call it an ontology; the term topic map is also used.
- When content managers or website information architects do it, they may call it a taxonomy, among other things.
- There is also a school of business rule modelers, who use "fact diagrams", which basically are concept maps supplemented with logic.

As you can see, the common denominator for all those communities of practice is: They are specifying *something they want to include in an IT-system* of some kind (many choices here). Logic is necessary and highly appropriate in that context. However, on the business level the world is often a bit fuzzy and we really do not need that extra precision, which logic adds to the picture. Simple business rules formulated in clear language is more than enough in most cases. Even business rules are possibly quite low level in the sense that they may exist with certain local variations, and they are subject to change more often than concepts and their relationships. From a business perspective what is stable is concepts and relationships. They survive most things, including change of ERP-systems, for example.

11.2 Business Rules Extend the Concept Maps

Even given that concept maps and business rules serve two different purposes, there is certainly synergy between concept maps and business rules, very much in deed.

First of all, business rules "fit into" the concept maps. From an overall perspective, all those "if ... then ... else ..." rules contain business concepts in the formulations of both the conditions and the consequences. Business rules cannot live without defined concept maps.

This is recognized by a school of business "rulers" working with "fact modeling". This is not actually completely new. Its predecessor was called Object-Role-Modeling (ORM), and it was supported by the popular Visio diagramming tool at the time that Microsoft bought the company behind Visio. See the book "Information Modeling and Relation Databases" (Halpin and Morgan 2008). ORM is definitely at the right level (concepts and their relationships) but it has built in all the logic details required for formal, precise specifications. This makes it (ORM) as complex as UML and in consequence it is not suited for business-level specifications (the visual syntax looks too complex for most business people).

Today devout modelers use fact modeling. Its biggest influence, however, has been as the platform for the new business rule standard called SBVR, (Semantics Of Business Vocabulary And Business Rules) from the Object Management Group standardization organization. It is a standard which came out in 2008, and its scope is defined as: "This specification defines the vocabulary and rules for documenting the semantics of business vocabularies, business facts, and business rules ..." (OMG 2008).

The SBVR documentation is building on the EU-Rent fictive case, of which there a few depictions (visualized as concept maps) in previous chapters.

The basic ideas of SBVR are:
- SBVR is expressed in a structured language close to written natural language.
- It can be used to define the business contexts, the business vocabularies and the detailed business rules.
- The "syntax" is essentially sets of predefined formulations such as "it is obligatory that ....", "each", "exactly one" and much more (which you have to learn and remember).

Our little pedagogical puppet kennel example could look something like this in structured English:
Puppy has *Puppy Number*
Puppy has *Puppy Name*
Kennel has *Kennel Code*
Kennel has *Kennel Name*
Kennel has *Kennel Location*
Kennel names *Puppy*
Trick has *Trick Name*
Trick has *Trick ID*
Trick categorizes *Puppy Trick*

11.2 Business Rules Extend the Concept Maps

Puppy Trick has *Skill Level*
It is possible that *Kennel* has for sale *Puppy*
It is possible that *Puppy* may do *Puppy Trick*
It is possible that *Where Learned* may be known of *Puppy Trick*

(The example above is deliberately pedagogical – its purpose is to illustrate the parallels between SBVR fact notation and rules versus concept mapping).

And this is the concept map underlying the SBVR-syntax above:

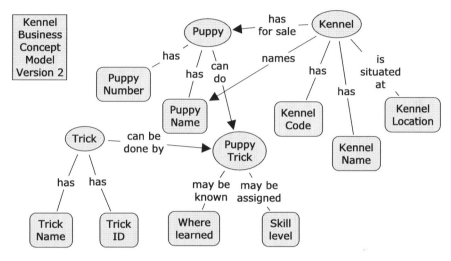

Fig 11.1 The kennel business concept map (revisited)

The implication is that SBVR is a formidable (technical) vehicle for taking business concept maps into the world of business rules. You have to express them as SBVR-statements. During that process a lot of detailed rules (logic) must be extracted from the (until now unstructured) descriptions of the business concepts and re-formulated as precise SBVR-syntax.

Why bother doing that, you might ask. There are some very good reasons for doing that. If you have a complex set of business rules, which also happen to change frequently, business rule automation is a definitive possibility for creating very good business value and a good return of investment in a relatively short period of time. The Business Rules Community website is a good place to start (www.brcommunity.com). What you see there will remind you of the recommendations of this book. The big difference is that you should divide and conquer:

1. Use business concept maps together with verbal definitions for most business-level descriptions (not more, not less than that) – this satisfies the **validity requirement.**
2. Use only all of the power of formal logic in business rules technologies only when it adds business value – this satisfies the **reliability requirement** of the day-to-day operations of the organization.

The basic mapping from concept maps to business rules fact models is trivial:

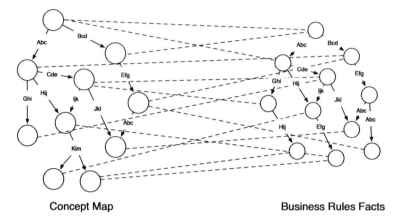

Fig. 11.2 Concept maps as stepping stones to business rules automation

If you want or need to get deeper into business rules, see. for example "Principles of the Business Rules Approach" (Ross 2003). A new and quite appealing approach is found in "The Decision Model – A Business Logic Framework Linking Business and Technology", (Von Halle and Goldberg 2010).

Opportunity: Reusable Business Information

12

One of the great challenges of the "information society" is that matching data is so difficult. Even in the traditional database world this is very costly. Think for example about all the programming efforts put into the data warehouse worlds extraction, transformation and loads of data.

Basically there are two levels of problems:

- Representational (e.g. how do you compare a decimal number with an integer), and
- Semantic/definitional (e.g. do customers, who have requested a quote, but not bought anything, count as "customers"?)

The problem scales up when you look at the internet. Try to Google for *software tool index* (looking for indexing tools), and you get something like this:

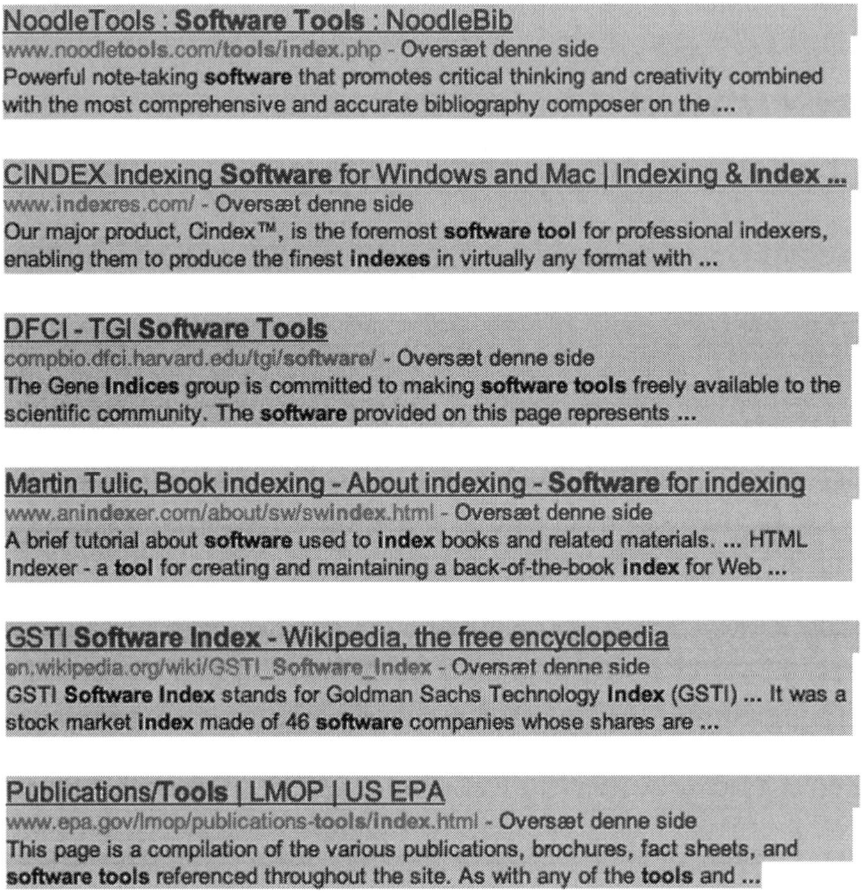

NoodleTools : Software Tools : NoodleBib
www.noodletools.com/tools/index.php - Oversæt denne side
Powerful note-taking **software** that promotes critical thinking and creativity combined with the most comprehensive and accurate bibliography composer on the ...

CINDEX Indexing Software for Windows and Mac | Indexing & Index ...
www.indexres.com/ - Oversæt denne side
Our major product, Cindex™, is the foremost **software tool** for professional indexers, enabling them to produce the finest **indexes** in virtually any format with ...

DFCI - TGI Software Tools
compbio.dfci.harvard.edu/tgi/software/ - Oversæt denne side
The Gene **Indices** group is committed to making **software tools** freely available to the scientific community. The **software** provided on this page represents ...

Martin Tulic, Book indexing - About indexing - Software for indexing
www.anindexer.com/about/sw/swindex.html - Oversæt denne side
A brief tutorial about **software** used to **index** books and related materials. ... HTML Indexer - a **tool** for creating and maintaining a back-of-the-book **index** for Web ...

GSTI Software Index - Wikipedia, the free encyclopedia
en.wikipedia.org/wiki/GSTI_Software_Index - Oversæt denne side
GSTI **Software Index** stands for Goldman Sachs Technology **Index** (GSTI) ... It was a stock market **index** made of 46 **software** companies whose shares are ...

Publications/Tools | LMOP | US EPA
www.epa.gov/lmop/publications-tools/index.html - Oversæt denne side
This page is a compilation of the various publications, brochures, fact sheets, and **software tools** referenced throughout the site. As with any of the **tools** and ...

Fig. 12.1 Google search result for "Software Tool Index"

Notice that some entries are there only because "index" is found in the URL (pretty normal thing). We also get both "Gene Indices" and Goldman Sachs Technology Index and "Cindex". The last one (Cindex) is probably the kind of answer; we were looking for to begin with.

It is for this reason that Google has introduced the next generation of Google search, called "knowledge graphs". It is explained on Googles website (www.google.com/insidesearch/features/search/knowledge.html) as – among other things – something that deals with meaning: "The words you search with can often have more than one meaning. With the knowledge graph we can understand the difference, and help you narrow your results to find just the answers you're looking for". When you look into it in more detail you find out that a "knowledge graph" is pretty much the same kind of structure as a concept map (supplemented with technical details, of course).

This is a good example of the increasing importance of automated support for "meaning". The paradigm is sometimes referred to as "Web 3.0" because much of

12 Opportunity: Reusable Business Information

the technology is based on what the original inventors call "The Semantic Web". This is a set of standards developed by the World Wide Web Consortium (W3C), the organization that also developed the specifications of HTTP, URIs and the XML-family of standards.

Today the semantic web environment is based on the following standards, all of which are supported by robust technology, in use by many large companies and organizations:

Acronym	Name	Purpose
SKOS	Simple Knowledge Organization System	Management of vocabularies of concepts and relationships
OWL	Web Ontology Language	Management of ontologies (structured vocabularies based on logic) with inferencing possibilities
SPARQL	SPARQL Protocol and RDF Query Language	A query language for RDF (databases)
RDF and RDF schema	Resource Description Framework	Definition and representation of concepts and relationships (the "data layer" of the semantic web)
XML and XML schema	Extensible Markup Language	The definitional platform for all of the above

If you want hardcore definitions of the above, go the website of the W3C consortium at www.w3c.org – otherwise you might be interested in reading a good book about it. For elaborate RDF and OWL and also SKOS guidance see: "Semantic Web for the Working Ontologist: Effective Modeling in RDFS and OWL" (Allemang and Hendler 2011).

OWL is by far the most feature-rich of the semantic web components. It is also sitting at the top of the stack. However, ontologies based on OWL are just as time-consuming to develop as UML (at a slightly lower level of complexity) and ORM/SBVR (cf. above). However, OWL is very powerful and has its own right as a component of complex solutions with very complex structures of meaning and sophisticated requirements for the precision (quality) of search results.

SKOS is the "light-weight" component for the definition of controlled vocabularies. It is based on the notion of concepts and their relationships, quite close to concept maps. We will get back to SKOS below.

Here are some facts about Woody Allen and some of his films:

Who	What	What
Woody Allen	Wrote	To Rome with love
Woody Allen	Wrote	Midnight in Paris
Woody Allen	Wrote	You will meet a tall dark stranger
Woody Allen	Wrote	Many more
Woody Allen	Acted in	To Rome with love
Woody Allen	Acted in	Scoop
Woody Allen	Acted in	Anything else
Woody Allen	Acted in	Many more
Woody Allen	Produced	What's up Tiger Lilly?

RDF is the "data layer" of the semantic web. RDF is based on the notion of triples as you can see below:

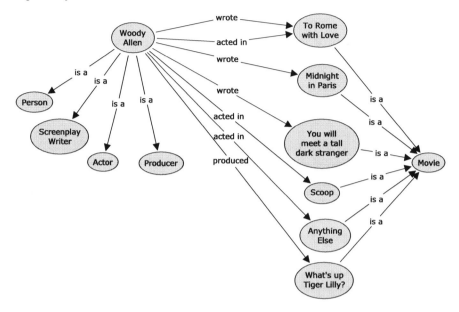

Fig. 12.2 RDF "graph" (simplified) of concepts related to Woody Allen

Each triple consists of "Subject", "Predicate" and "Object" (Subject: Woody Allan, Predicate: wrote, Object: Midnight in Paris). And it may be read as a simple sentence. Just like the concept-relationship-concept combination in concept maps. In fact the combination of many triples becomes a graph resembling a concept map. (The diagram above is not a concept map in the sense that is used in this book, but the similarity is striking).

Here is the "complete model" of a Movie database for comparison:

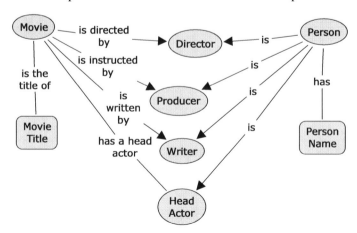

Fig. 12.3 RDF style model of a movie (graph) database

12 Opportunity: Reusable Business Information

RDF may be stored in specialized databases, of which there are plenty ("RDF stores", "Graph databases" and more), or RDF may be provided as an interface to a database (either as part of a database product or as a "bridge" to a database product). SPARQL is the query language (quite technical) that comes with RDF.

The basic idea of meaningful reuse consists of two parts:
- Data is presented in a standard format (RDF)
- Data are described by way of a standard (SKOS or OWL).

The important component from the perspective of this book is SKOS. Basically SKOS can be used to represent concept maps. SKOS has slightly more to offer. SKOS is a very simple starting point. It basically supports both naming concepts and specifying relationships between concepts. There are two types of structural relationships:
- Hierarchical – expressed as *broader* or *narrower*
- Related on the same level (simply called *related*)

In this diagram you see a simple SKOS construct (represented as a concept map):

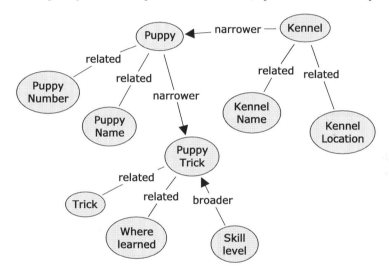

Fig. 12.4 The kennel concept map using SKOS terminology

As you can see there is no specific name on the relationships. SKOS is meant as an easy entry-point into the world of the semantic web standards. That means that you could (if you know the syntax) extend your SKOS-model with. e.g. the "subPropertyOf" or the "subClassOf" mechanisms found in the RDF Schema "toolbox". Once you go down that road you will find that you have opened up a treasure chest of possibilities.

In consequence it is very easy to get from concept maps to SKOS and RDF. And from SKOS and RDF you can go to RDF to OWL by adding logic.

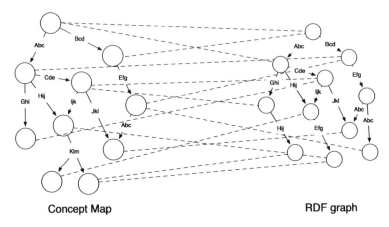

Fig. 12.5 Concept maps map to RDF graphs (and SKOS)

Why is that fantastic? First of all there is the "meaningful" results, which we discussed in the start of this chapter. That is, of course, very useful for searching. In addition to that it greatly facilitates:

- Integration between structured data (databases such as data warehouse) and unstructured data (e.g. documents on an intranet or in a CMS-system)
- Meaningful business intelligence (as described above), but possibly also supported by semantics-aware software
- Easier integration between in-house databases in an organization
- Integration with external data and information; see the next chapter.

The semantic web technologies have matured the last 10 years. And they are now in production use in many large companies such as Microsoft (Bing) and Google and many more. Your organization should also start to take advantage of semantics. And the place to start is business concept mapping, which forms the core of the business foundation of developing the business using new (IT) opportunities.

Remember, concept maps address the business validity issue, whereas sophisticated semantics (like OWL) support the operational reliability of complex IT solutions. You cannot get to reliability without a good start in validity.

Opportunity: Open Information Sharing 13

Data, which are available in RDF-format and are described using SKOS/OWL (cf. above), can be made available not only internally in an organization, but also externally. You add some extra (technical) protocols etc. and you have what today is called Linked Data or Linked Open Data.

This is a fast growing movement on the internet. Take a look at http://linkeddata.org. At the time of writing there are a little less than 5,000 registered sites offering linked data – a number of which are open. There are very many different categories available; too many to list here.

Some of the most known are:
- DBpedia (database version of Wikipedia)
- Freebase (now part of Google – is an open source type of data collection)
- Geography and demographics
- Media data
- Government data
- Libraries and universities
- Life sciences
- Business information
- Retail and commerce
- Social media

The Wiki people are busy working on a new free service called WikiData. It will contain open, described and structured data from all over the world (in many languages). Just like Wikipedia – only now machine-readable with semantic integrity.

Many people describe the Linked Data approach as the internet done right. And it is true that the vision of a global data space goes back to the earliest days of the Internet.

What is in it for my organization, you might ask? Obviously there is a lot of data, which you need from the outside. Using the semantic web technologies (OWL, RDF and so forth) you can now much easier integrate external data with your own. No need for many of the data integration projects that goes on in all those data warehouses and elsewhere. That is a significant advantage in terms of agility, flexibility and business quality.

The trick is to describe your own business information in a way that makes it easy to match them with incoming data described using semantic web standards. As described in the previous chapter this is a matter of taking your business concept maps and translate them to either SKOS or OWL with the necessary enhancements. You do the mapping on the business level, basically as a one-time task – not as a per project mapping task. In short the picture is the same as for the other semantic innovation opportunities:

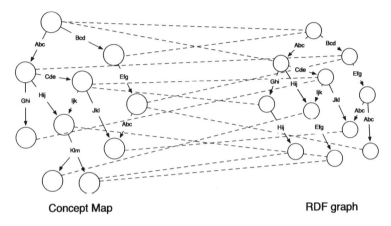

Fig. 13.1 Concept maps and RDF graphs (revisited)

Your business concept maps thus make it easier to use open data for business innovation. Because concept mapping is on the same level of structure (concepts and relationships) as are SKOS and RDF.

14 Opportunity: Pull Instead of Push

There is one more benefit of working on the business concept level. As described in the previous chapters mapping the concept maps to SKOS and RDF standards makes it easy for you to integrate your data with open data from the outside.

The opposite way is just as easy. What that means is that internet users (some of which are customers/clients/citizens/business partners and so forth) can intelligently browse your data with the same high precision as for any other web 3.0 search. They get only the kind of results they meant to ask for, not the useless hits of yesterdays search engine technology. The current state of affairs is, consequently, that – because of the tediousness of browsing search results – people go to specialized sites for buying a house, a car or a kitchen table (or some T-shirts manufactured in low-cost production sites overseas).

What that means is that today sellers must push marketing information to perspective clients. Resulting in information overload, spam and consumers developing marketing tolerance (not seeing or listening) or marketing intolerance (shut down the TV or leave the site).

In his excellent book: "Pull, The Power of the Semantic Web to Transform Your Business" (Siegel 2009), David Siegel points out that there is another way. What if I (as a potential buyer) describe my needs and what I have using a defined structure (in other words based on semantic web technology)? Then sellers could pull out (by way of semantic search) a prospect list of only relevant potential customers and talk to them directly. So the potential buyers advertise their potential requirements instead of the seller advertising their unique selling points to everybody (most of whom are not interested, at least most of the time).

Again, this is a semantic innovation opportunity and it looks like this:

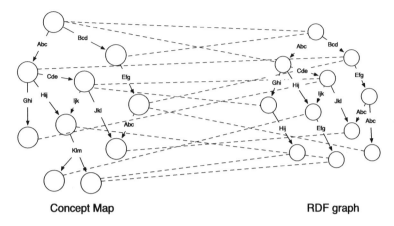

Fig. 14.1 Concept maps and RDF graphs (revisited)

The interesting thing in the context of this book is that only by using business concept mapping is the business organization equipped to deal with the business opportunities created by semantic web technologies.

Opportunity: NoSQL and Big Data 15

A lot of attention is being given to new ways of dealing with very large amounts of data. For some (e.g. the very big sites like Amazon, Google and so forth) SQL-databases suffer from both inflexibility and lack of performance on the terabyte level. Hence the NoSQL-movement centers on a "big table" concept as in HADOOP, Hive and other technologies. Essentially the data model is a hierarchical model of grouped columns. Some columns may point (using URIs) to rows in other tables.

But there are other approaches.

In the financial sector, primarily, a fresh approach to data modeling emerged some years ago. Under the names of "Data Vault Modeling" (a proprietary method by Dan Linstedt) and "Anchor Modeling" (an open source methodology), the modelers went to the ultimate normalization level (sixth normal form for the nerds among the readers). What this essentially boils down to is that, instead of having tables with columns, each column becomes something for itself – having relationships with other somethings. Those "somethings" are, of course implementations of business concepts (and some technical attributes, maybe).

A new family of database products uses the column paradigm. Instead of storing rows of tables, they store (compressed) occurrences of values in columns in tables. (With some smart mapping to make it easy to present the data as rows).

Finally there is also an emerging group of "graph databases". Some of them are RDF databases (RDF is essentially a graph paradigm), where others are not at RDF yet. As we have seen previously there is an almost one-to-one correspondence between concept maps and RDF.

What you can see here, is that the IT data model paradigms shift away from tables, columns and foreign keys (of the classical data modeling schools). The beneficiary is a model focused on concepts (columns) and relationships. The distance from a concept map to a database model narrows down considerably:

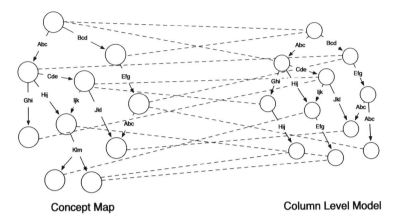

Fig. 15.1 Concept maps front-end column level data models

As it happens, column level data models are most often quite technical. Again, you need to map from the technical naming schemes to the business level terminology defined by the business concept maps.

Think Big, Start Small: Deliver Value to the Business

16.1 Simple Tools that Work

One of the leading thinkers within information modeling, Malcolm D. Chisholm has stated:

> Conceptual models must capture all business concepts and all relevant relationships. If instances of things are also part of the business reality, they must be captured too. Unfortunately, there is no standard methodology and notation to do this. Conceptual models that communicate business reality effectively require some degree of artistic imagination. They are products of analysis, not of design. (Chisholm 2012)

The key words here are: Concepts (what are we talking about), relationships (between concepts; customers place orders and so forth) and "things" (Product XYZ, Customer Thomas etc.). This is the realm of the business, not of IT. Do not use engineering tools here – time has shown they do not work!

The important thing is to get the job started, in order to be prepared for taking advantage of a better business understanding and of the business semantics. Emphasis should be on taking control of this challenge and turn it into business development opportunities. And on working with IT on the technical matters such as semantic technologies and text analytics, for instance.

16.2 Tested Approach

Based on experiences from many projects since 2005 (and ongoing), there is strong confirmation of the applicability of concept mapping for business analysis. Most of the projects have been data warehouse/business intelligence oriented business development projects. A few of the other projects were large specification projects in the public sector.

There is positive confirmation of:
- Concept maps are indeed intuitively easy to understand for business people, which greatly simplifies and facilities more agile business analysis activities.

- Concept maps are well suited for collective brainstorming sessions (workshops) where the concept maps are being drawn (on a computer connected to a data projector) as the discussion goes on. (IHMC CmapTools is used for this).
- Concept maps can be reviewed, maintained and enhanced by business people (with some guidance).
- Design thinking and concept mapping support each other extremely well.

A simple visual "syntax" involving "Business Objects", "Properties" and "Relationships" has proven to be very effective in communicating information structure to business people. Client feedback has been overwhelmingly positive. The business users quickly recognize the added value of concept maps. The viability of "meaningful learning" (which is the theoretical background of concept mapping) is solidly confirmed.

On the business level the world is often a bit fuzzier than IT people really like. We really do not need that extra precision, which logic-based approaches (such as UML or OWL) add to the picture. Simple business rules formulated in clear language are more than enough in most cases. Even business rules are possibly quite low level in the sense that they may exist with certain local variations, and they are subject to change more often than concepts and their relationships. From a business perspective what is stable is concepts and relationships. They survive most events, including change of ERP-systems, for example.

16.3 Benefits of Business Concept Modeling

Time spent in the early days of the project on business concept modeling is time well spent. You will, actually, gain in the end, and the whole of the project is more safe and manageable. This is because:
- You are certain that the end result, e.g. a BI solution, is intuitively accessible for business users.
- You understand the business 100%.
- The business learns hidden facts about itself (many "Aha!" experiences on the way).

Many people expect that the business concepts and their structures are well known to everybody. That is not the case. Somebody in the business organization must begin to manage this valuable asset.

There is a direct path from a concept map into a multidimensional environment with dimensions and hierarchies etc. For example setting up the UDM (Unified Dimensional Model) of a simple one-to-one process, because all the attribute relationships are well defined and understood already.

The design thinking style approach using concept mapping has repeatedly produced very valid designs of business information at the level of concepts and relationships. The acceptance by the business community is much higher than anything else, we have seen. The effect of producing more valid business information structures is, of course, a platform for more reliable business management.

What has happened in traditional business analysis is that the requirement of reliability has overshadowed everything else, maybe because IT for many years was still "new" and slightly dangerous technologies. This lead to people (including many business analysts) to prefer "engineering style" approaches (like e.g. UML class diagrams) to deal with business issue. But that is a wrong tool in a wrong place. Dealing with business issues on the tactical/strategical development scale is driven by (business) validity, which requires somewhat different approaches – definitely not rigid engineering methodologies. Something more agile and inventive is necessary.

In order to change something, you must understand it. That is why your concept maps are to the business information asset what the chart of accounts is to the financial assets.

16.4 Design Thinking

Concept mapping is by itself of great value to the organization. It is when you use it to support design thinking approaches to business development that you really produce new business value. The nature of business analysis is closely related to the nature of what a designer or an architect does. Picture the enterprise as a building:

Fig. 16.1 An "introvert" office building in Copenhagen, Denmark (Photograph by the author)

In the beginning the task looks immense. You look at a very complex structure, much like a huge building with many floors, endless corridors and more doors than you can count. Some of the doors may be hidden, secret or forgotten. But you need to know what is behind them, in due course.

Fig. 16.2 An office corridor at a University in Chennai, India (Original color photograph by Pradeek Karandikar, 2009, available on Wikimedia Creative Commons)

Much like the architect would do, you start walking around and examine the floors, the staircases and so on. You take snapshots (concept maps) on your way. It takes time and you need to speak with many people, some of which are very busy. You start to draw an overall "floor plan" (concept map) for each floor.

Having established an overview, you and the business people select some area, which badly needs to be refurbished. You measure it up and document your findings in one or more detailed plans (concept maps). Together with the business people, you start considering alternatives – sketching them (as concept maps) as you proceed. You build a model (concept map or prototype) or two – "proof of concept" in a very literal sense.

It is a difficult process – there are many alternatives. The purpose of the whole exercise is to create real, new business value by way of innovation. The optimal solution could be to reduce throughput time by making it easier and faster to come from one place to another.

Finally a decision is reached (e.g. "we need a new staircase from A to B"). The implementation details are specified (in a concept map) and you call the "masons" and the "carpenters".

Once finished everybody agrees it is a good idea. "Why didn't we think about this before?" The people involved thank the business analyst for the insightful assistance.

16.5 Summary of Guidelines

Success leads to the next success. The design thinking approach combined with concept mapping is powerful, productive, creative and repeatable.

Here is a brief listing of the most important guidelines:
- Establish an overview early on (using high-level concept maps of business objects).
- Break down the high-level concepts in small portions to be analyzed according to the business priorities and business development plans.
- Unless you want to get into business rules automation you should document the more important (and long-lasting) business rules as part of the text definitions document.
- Spread the knowledge in the organization about the concepts, their structures and their definitions – everybody will benefit from "speaking the same language".
- Organize concept maps top-down (the higher-level business objects in the top).
- If there are implicit "logical" successors from a business events point of view, organize the business objects from left to right (e.g. budget before general ledger and so forth). Do not use arrows to indicate succession!
- Do not mix process flows into the concept maps – concepts maps are about the structure of meaning, not about processes and flows.
- Keep the focus – do not mix exploration, ideation and implementation; take one at a time (even if each of the three stages may well be very short periods of time).
- Strengthen the creativity of the ideation stage by way of massive, alternative sketching (concept maps) possibly supplemented with rapid prototypes based on live data (you need a good data person for this).
- Bear in mind that there are at least six different types of concept maps: high-level overviews, explorative as-is/wannabe, prototype/brainstorming sketches, working prototype designs, solution design (depending on target platform etc.), documentation/instructional.
- Remember to challenge established "facts" – chances are you will discover hidden truths and unspoken misunderstandings.
- Think What, Where, Who, When and How much.
- Look for the real business pains.
- Reliability is what makes the business manageable.
- Validity by design of business concepts, process, rules etc. is what enables reliability. Validity is the driver of business analysis and business development.
- If in doubt, start at the very top – the business model.
- Use a simple visual syntax in your concept maps (you may design your own or you can use the one suggested here).
- Do not put more than 20 concepts on a concept map (use an "off-page" notation).
- If you have a data modeling background, please be very aware of the fact that concept mapping is not data modeling (except for some of the implementation stage concept maps, maybe).

- Rely on brainstorming workshops as the most robust source of knowledge about business concepts, structures and definitions; supplement as necessary with knowledge collected from Excel-sheets, databases and so forth.
- Keep the business development opportunities using new technologies (as suggested in this book) in the back of your head.
- Have fun and enjoy the creativity in good company!

To sum it up: The roles and tasks of business analysts are changing, because they transform into "Business Synthesists" (using design thinking and concept maps). This way validity becomes the key driver because it goes hand in hand with innovation.

Remember this wherever you go:

What is of business value belongs to the business!

Appendix 1: Explanation of Terms and Acronyms

Term/acronym	Meaning
Big data	Used to describe a number of disciplines of data management and analysis of very large volumes of data (terabytes and larger). Often connected to NoSQL (cf. below)
Business object	Loosely defined as both physical and abstract "things" found in the business terminology such as customers, invoices and so forth
Business rule	Detailed logic describing some very specific conditions, which must be satisfied in a given business context. E.g.: A rental car, which is a SUV, is always placed in rate class E
Business rule automation	Software packages, which from a collection of business rules (expressed in some form of formal logic), makes inferences about an outcome. E.g.: In a given situation, rate class is B
Concept	Here loosely defined as a denominator (name) of a type of "thing" or properties of "things" (cf. business object above). E.g. "customer", "rate class" and so forth
Concept map	A visual representation of concepts and the relationships between them
Content extraction	Software packages, which are capable of extracting descriptions of the content of documents etc. Typically used for classification of the documents
Data profiling	Software packages, which are able to perform a broad selection of measures and tests on the contents of a database or a file. The resulting data profile is used to analyze potential risks in converting or extracting the data. (In order to scope and size the data conversion project or to build quality measures about the state of a given database)
EIM	Enterprise Information Management (EIM) is the term used today for management of information and metadata at the enterprise level
ERP-system	Enterprise Resource Planning (ERP) software package. Used today to denote many different kinds of both general and industry-specific application solutions
EU-Rent	EU-Rent is the name of a fictitious company used in the OMG SBVR-specification (see below) for the examples in the documentation
Graphs	In the context of this book "graph" means a directed graph such as a concept map. The RDF-standard (cf. below) is also based on graphs. There are software packages, which support graph databases directly
Hierarchy management	Hierarchies are important in business intelligence. They are used to "drill-down" respectively "roll-up". E.g.: Year-Quarter-Month-Day is a hierarchy. Hierarchy management is found in business intelligence software as well as in master data management (MDM cf. below) software

(continued)

Term/acronym	Meaning
KPI	Key Performance Indicators (KPI's) originate from the balanced scorecard paradigm and indicate central measures, which the business organization follows closely (on a "scorecard"). Today the term is used more loosely to denote important key numbers for the management of an organization
MDM	Master Data Management (MDM) is a set of best-practice guidelines and procedures for managing important, shared data in an enterprise. MDM may be supported by specialized software packages
Measure	Used in business intelligence jargon to designate facts about the performance of a business process. E.g.: No of products sold and so on
Multidimensional	Multidimensional models are used in business intelligence. Within the paradigm a set of measures is presented as if they were cells in a multidimensional "cube". The paradigm is intuitively easy to work with and it is supported by software packages. Each "side" of the "cube" represents a dimension – within which you may find hierarchies
Normalization	A technique used by data modelers to design the structure of the database. The rule is that any field in a database can be in only one place and that is in the table, which is uniquely identifying the business object in design. Normalization is related to hierarchies in that the hierarchies will result in a set of tables, on for each level in the hierarchy
NoSQL	Used very loosely to denote a number of different technologies and approaches to data management. The defining characteristic is that the technology is not based on a SQL-database (!) The term is very often connected to Big Data (cf. above) and really comprises a number of very different data management schemes, such as hierarchical, self-describing data, graph databases and column stores (and much more)
OMG	The Object Management Group (OMG) is an international standardization body, which has produced a number of important standards related to development methodology etc. Cf. www.omg.org
Ontology	A term used in knowledge management and the semantic web (cf. below) paradigms. It denotes a facility for building highly structured "dictionaries" based on formal logic. OWL (cf. below) is a standard for ontology descriptions
OWL	A standard from the W3C consortium (cf. below) for specifying ontologies. It is based on formal logic and enables inference
RASCI	A method for describing relations between people and a project (or parts thereof). The letters stand for: Responsible, accountable, supportive, consulted and informed
RDF	Resource Description Framework (RDF). Definitions and representations of concepts and relationships (the "data layer" of the semantic web (cf. below). It consists of schema facility (for defining a model) and the data facility for representing instances of graphs
Relationship	In the context of this book "relationship" means relationships between concepts. In concept maps relationships are named. E.g.: Customer – places – Order, where "places" is the relationship
SBVR	Semantics Of Business Vocabulary And Business Rules (SBVR). A business rules standard language from the OMG (cf. above). There is a "bridge" defined between SBVR and OWL (cf. above)

(continued)

Appendix 1: Explanation of Terms and Acronyms

Term/acronym	Meaning
Semantic web	A project within the W3C consortium (cf. below) to produce standards for supporting semantics (cf. below) and thus enable much more intelligent searches with highly relevant result sets. Some of the semantic web standards are URI, XML, RDF, OWL, SKOS and more. Cf. the individual entries for each of the acronyms
Semantics	Semantics is dealing with representation of meaning. Defined semantics is useful for all communication – be it man-machine or machine-machine. Concept maps is an example of defined semantics. An OWL-specification is also a (much more detailed) example of defined semantics
Sentiment analysis	A term used in text analytics (cf. below). Software packages exist, which can extract the "sentiment" from a document. E.g.: Extracting information about whether a customer is annoyed based on an email received from that customer
SKOS	Simple Knowledge Organization System. Management of vocabularies of concepts and relationships. A standard within the semantic web family (cf. above)
SPARQL	SPARQL Protocol and RDF Query Language (self-referencing!). A query language for RDF-databases. A standard within the semantic web family (cf. above)
Star schema	A data modeling technique for representing multidimensional data (cf. above) in relational databases
SWOT	An analysis technique for looking at alternative strategies and plans. The letters stand for Strengths, Weaknesses, Opportunities and Threats
Taxonomi	A model for representing hierarchical classification systems. Originally used in biology and library science. Today more broadly applied across many different usage types (including e.g. website information architecture)
Text Analytics	Software packages, which analyze text in order to e.g. arrive at classification schemes and many other types of information from the documents
Text Mining	Text mining is an older term, which is being replaced by text analytics (cf. above)
Thesaurus	Essentially a library science term for a structured vocabulary
UDM	A Microsoft SQL Server model for defining multidimensional structures. Includes "attribute relationships" for building hierarchies and it is – because of that – very close to the concept mapping level
UML	Unified Modeling Language (UML) is an OMG (cf. above) standard for specification of IT-systems and related development information. There are many visual components, not least UML Class diagrams
URI	Uniform Resource Identifier (URI). Is a standard for specification of a resource by name (URN) and location (URL). URL's are the well-known web-addresses. A standard within the semantic web family
URL	Is the location-form of a URI, cf. above
W3C	The World Wide Web Consortium. An international standards organization developing the standards for the internet
Web 3.0	Referring to the same as the "semantic web", cf. above
XML	Extensible Markup Language. The "definitional platform" for all of the semantic web standards (cf. above)

Appendix 2: User Guide to CmapTools

How to Use CmapTools

The tool we use for drawing the concept maps is called CmapTools. It is based on theoretical work done in the eighties and the nineties by Professor Joseph D. Novak (at that time at Cornell University), cf. (Novak 2008). The tool, CmapTools as we know it today, came to life around year 2000 and it picked up speed in the educational sector a couple of years later. Today the publisher of the software is the Florida Institute of Human Machine Communication, IHMC, where Prof. Novak works.

The tool, which is free, can be downloaded from its website at http://cmap.ihmc.us. It is intuitively easy to use and is ideally suited for brainstorming sessions around a data projector. There are plenty of sources of guidance to CmapTools on the Internet, not least on IHMC's website: http://cmap.ihmc.us/support/help/. But alternatives do exist – find them with Google and YouTube. (CmapTools is quite popular in Spanish speaking countries, so you will find a number of hits in Spanish).

This is not another "User Guide to CmapTools", but it is rather some observations, which have been found to be useful. Read them, if you are going to download and install CmapTools and start using it.

After the installation you will be asked for a username and password. That is because CmapTools can be used together with Cmap-servers. Either your own or the public ones provided by IHMC. But CMAP can be used just fine as a local program on your own computer. The default location of the concept map files (xxxx.cmap) is in a folder called My Cmaps in your Documents folder.

Take time – maybe a half-hour or more to play with the tool and try to recreate one or more of the concept maps in this book.

Use the Style Palette Window to manipulate the objects in the drawing. Highlight something, and you can change the appearances of text, lines and objects. It is also here, in the Style Palette, that you add or remove arrowheads. Play with it a couple of minutes till you get to know it better.

Color is frequently used to highlight concepts, which are "work in progress". I.e. they can be things you are not sure of, need to investigate or need to decide on. Yellow (like a Post-It) works fine for most people. (Use the style palette with the concept highlighted to change the color of the object).

Note that CmapTools can actually do very many things, including linking to other resources (e.g. web pages) and so. If you want to get into that, consult the help page at IHMC (cf. above).

If you want to include a concept map in e.g. a Word document or a Powerpoint, the easiest way to do that is to "Export Cmap as Image File" in the menu. Use e.g. EPS-format and it works fine, once you insert the picture file into your document.

Here is a short description of the syntax, which is proposed for you to use for business concept mapping:

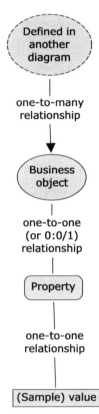

Fig. A.2.1 The visual syntax proposed for business concept mapping

Round icons symbolize business objects, which can include physical objects (e.g. goods), documents (e.g. invoices, etc.), actors (people/organizations), events/transactions. (E.g. sales, posting, distribution, etc.)

Square icons with rounded corners symbolize the characteristics (properties) of business objects, e.g. Product Weight, etc.

Square icons with sharp corners are used occasionally to exemplify one or more values of a property (e.g. "British Racing Green" or "Red"). This is purely for pedagogical reasons.

Connecting lines indicates relationships between concepts. If the connection is made with an arrowhead, it indicates a one-to-many relationship (e.g. multiple passengers per trip). Without the arrowhead the meaning is that there is either a 1-1 or a 0-1 relationship. In both cases it may also be that there is not always a relationship, for example, there may be a situation where a new Customer not yet has bought anything. These variations are expressed using, for example "has" respectively "may" or similar as the text of the relationship.

This particular way of using CmapTools (and concept mapping) in general is just one way of doing it. Feel free to create your own style. What matters is that the concept maps *communicate meaning well to business people*. Remember to work from the top and down and from the left to the right, when you layout a diagram.

Appendix 3: Comparison of the Concept Level and the Entity Level of Modeling

Business Information Modeling

Business information modeling is one of the most overlooked and under-supported areas of business development using IT.

The importance of business information modeling becomes very clear in business intelligence (BI) and data warehousing. Think of it: What is business intelligence first and foremost? The answer is: A presentation of business information intended for direct access by untrained business users, who knows the business well, but who do not know a whole lot about information technology.

Since the deliverables of BI and DW development projects are databases (and multidimensional cubes) there are next to none application programs, where IT developers can hide business rules and other logic.

A successful business information model should have at least these important characteristics:
- It is intuitively understandable by business users.
- It expresses itself in the language of the business.
- It contains all the necessary business concepts and business rules.
- It is well defined and precise without redundancy and inconsistencies.
- It makes scoping and sizing the project much easier.

The above is the scope for what should be delivered by a business analyst as part of a business development project (be it business intelligence, semantics or more mundane ambitions such as process renovation and the like).

Business Metadata

One way of looking at this is as a way of collecting metadata (information about the way business works). Bill Inmon and Bonnie O'Neill et al. have written a very good book about this subject (Inmon et al. 2008). Business concept modeling is indeed about collecting metadata. However, "metadata" is a "what are you talking about" notion to most business people. So, please do not call it that.

Modeling Tools

By tradition IT people consider Entity-Relationship models or UML class diagrams, when they select a tool for modeling. Even if what they are asked to do is a conceptual model. But both approaches are too complex for business users, who really ought to take part in the modeling. As this book documents a simple tool for concept mapping is highly successful with the business users. We have successfully employed the product CmapTools for a number of years now. And we can highly recommend it, because it is intuitive, simple and powerful. And the concept maps work for business people.

Where Does the Complexity Come from?

When you attempt to go beyond the simplicity of a good concept map, you are essentially heading into the same problems discussed earlier in this book about adding logic. UML, for instance, is an almost complete solution to logic-based modeling.

However, UML and ER-diagrams are not the only paradigms in modeling. There are different forms and styles for different usages and/or different communities:
- Data models (IT)
- Taxonomies, facet plans or ontologies (library scientists/archivists)
- Topic maps or ontologies (knowledge managers)
- Taxonomies (content managers or website information architects)
- Fact diagrams (business rule modelers)

As you can see, the common denominator for all those communities of practice is: They are specifying *something they want to include in an IT-system* of some kind (many choices here).

No UML, Please!

Let us inspect some of the reasons that boxes and arrows as well as the rigid, but powerful, representations of e.g. UML or even ER-diagrams, are not suited for Business Concept Models.

Appendix 3: Comparison of the Concept Level and the Entity Level of Modeling

Let us go back to the kennel business concept map:

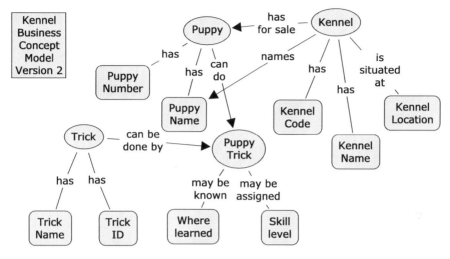

Fig. A.3.1 The kennel business concept model (revisited)

If you translate the above to a "Boxes and Arrows" type of diagram, you will experience information loss. Here follows an (almost) equivalent representation of the kennel concepts expressed as a generic UML-style class diagram (keys hidden):

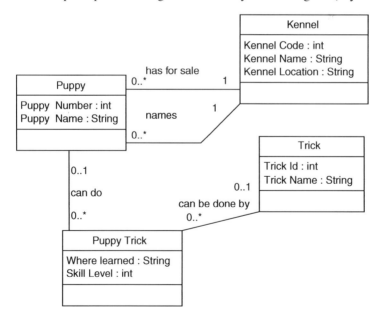

Fig. A.3.2 The kennel business concepts as a UML class model

As you can see, there are a number of instances of information loss, some of which are:
1. "Puppy Trick may be assigned Skill Level" becomes "Puppy Trick has Skill Level".
2. "Kennel is situated at Kennel Location" becomes "Kennel has Kennel Location".
3. Puppy Number and Puppy Name: the dependancies from Kennel are unclear (unless the data modeler names the foreign keys fields properly).

Number 1 would require a NULL-specification in a database since there is no optionality between the primary key and a functionally dependent attribute. But even that would not make it correct, because the original business terminology is lost. And lost means lost – it is not there – users have to guess what the implied functional dependency means. This is the most common problem – specific vocabulary degenerating to "has".

Number 2 has the problem that Kennel Location furthermore probably is a subtype of a generic Location concept, which is not in the model (yet?). In the diagram above it looks more like a design decision not to model Location in detail, but just to provide a long text field for users to enter the location into. But that is really a solution level design decision. What we want to model is the reality of the concepts of the business.

Number 3 is also confusing to users because the names of the relationships are "lost in translation" to the data model, and it could well be important for the business.

Some of the potential problems are visualized above – here is a more comprehensive (and technical) list:
- Loss on inter-attribute relationship information
- Object or attribute – instead of just concepts
- Multi-valued attributes are difficult to express in simple manner
- Roles represented as attributes – how to depict uniqueness, exclusion/inclusion etc.
- Object vs. association – often leads to new concepts and business keys (previously unknown to the business).

The biggest problem, however, is complexity. For the purpose of business concept modeling the mantra is: Less is more! Intuition is obstructed by unnecessary information. UML class diagrams offer (at least) meta-constructs such as:
- Association vs. aggregation vs. composition
- Generalization, sub-typing and realization
- Inheritance
- Multiplicity (cardinalities)
- Data types etc.

All of the above is not really necessary for business users to understand. Obviously one answer could be to suppress/avoid using all of it in some simplistic class diagrams (if the diagramming tools permits that). But then, why use UML in the first place? Simple Microsoft PowerPoint or Visio diagrams could be sufficient. But concept maps are better and have proven to work.

Further down the production line (in the design and specification phases) there may well be good reasons to use UML, depending on the circumstances.

The schism, which is behind this, is between purposes/requirements:

(A) Business quality understood as relevance and closeness to the business needs – what Roger Martin (Martin 2009) calls validity – and
(B) Operational engineering quality in the sense of producing reliable results in a complex world (Roger Martin's reliability dimension).

The business itself is responsible for (A) above and needs a tool to help them (concept mapping). IT/knowledge engineers and so forth are responsible for (B) and have plenty of tools already for dealing with the complexities of the resulting solutions.

References

Allemang D, Hendler J (2011) Semantic Web for the working ontologist: effective modeling in RDFS and OWL. Morgan-Kaufmann, Waltham

Aminoff C, Hänninen T, Kämäräinen M, Loiske J (2010) The changed role of design. commissioned by the Finnish ministry of employment and the economy to provoke design Oy/Ltd: www.tem.fi/files/26881/The_Changed_Role_of_Design.pdf. Accessed 29 Jan 2012

Brown T (2008) Design thinking. In: Harvard business review (hbr.org), June 2008

Cadle J, Paul D, Turner P (2010) Business analysis techniques – 72 essential tools for success. BISL (BSC), Swindon

Chisholm MD (2010) Definitions in information management – a guide to the fundamental semantic metadata. Design Media, Port Perry

Chisholm MD (2012) Big data and the coming conceptual model revolution. http://www.information-management.com/newsletters/data-model-conceptual-big-data-Chisholm-10022303-1.html. Accessed 7 Sept. 2012

English L (2009) Information quality applied. Wiley, Indianapolis

Gärdenfors P (2000/2004) Conceptual spaces. MIT Press, Cambridge

Halpin T, Morgan T (2008) Information modeling and relation databases. Morgan Kaufmann, Burlington

Inmon W, O'Neill B, Fryman L (2008) Capturing enterprise knowledge – business metadata. Morgan Kaufmann, Burlington

Jung CG (1993) Synchronicity: an acausal connecting principle. Bollingen Foundation

Kimball R (1996) The data warehouse toolkit. Wiley, New York

Ladley J (2010) Making EIM (enterprise information management) work for business. Morgan Kaufmann, Burlington

Laney D (2012) Introducing infonomics: valuing information as a corporate asset. Gartner Group research note G00227057

Liedtka J, Ogilvie T (2011) Designing for growth – a design thinking toolkit for managers. Columbia University Press, New York (Kindle edition)

Loshin D (2009) Master data management. Elsevier, Burlington

Martin R (2006) Design thinking and how it will change management education: an interview and discussion. Acad Manag Learn Educ 5(4):512–523

Martin R (2009) The design of business – why design thinking is the next competitive advantage. Harvard Business Press, Boston (Kindle edition)

Moon BM, Hoffman RR, Novak JD, Cañas AJ (2011) Applied concept mapping – capturing, analyzing and organizing knowledge. CRC Press, Boca Raton

Novak JD (1990) Concept maps and Vee diagrams: two metacognitive tools for science and mathematics education. Instr Sci 19:29–52

Novak JD (2008) Learning, creating, and using knowledge: concept maps® as facilitative tools in schools and corporations. Routledge, New York (Kindle edition)

Novak JD, Cañas AJ (2006) The theory underlying concept maps and how to construct and use them. Technical report IHMC CmapTools 2006–01 Rev 2008–01 – downloaded from http://cmap.ihmc.us/publications/researchpapers/theorycmaps/theoryunderlyingconceptmaps.htm#_ftn1

OMG (2008) Semantics of business vocabulary and business rules (SBVR), v1.0. OMG Document Number: formal/2008-01-02, http://www.omg.org/spec/SBVR/1.0/PDF

Osterwalder A, Pigneur Y (2010) Business model generation. Wiley, Hoboken

Oxman R (2004) Think-maps; teaching design thinking in design education. Design Stud 25(1):63–91

Ross R (2003) Principles of the business rules approach. Addison-Wesley, Boston

Siegel D (2009) Pull, the power of the semantic Web to transform your business. Portfolio/Penguin, New York

Sullivan W, Rees J (2008) Clean language – revealing metaphors and opening minds. Crown House Publishing, Carmarthen

Von Halle B, Goldberg L (2010) The decision model – a business logic framework linking business and technology. Taylor & Francis, Boca Raton

Made in the USA
San Bernardino, CA
28 August 2017